普通高等教育教材

化学工业出版社"十四五"普通高等教育规划教材

环境化学实验
Experiments in Environmental Chemistry

（双语）

弓爱君　主编　　邱丽娜　张玮玮　副主编

化学工业出版社

·北京·

内容简介

《环境化学实验》双语教材为高等院校"环境化学"相关课程的配套教材。本书共四章，第一章绪论，主要内容包括环境化学实验的目的和任务、实验室安全守则和实验室意外事故处理；第二章常用仪器使用方法，主要内容包括玻璃仪器的洗涤及干燥方法、环境化学实验用水及制备和环境化学实验常用仪器；第三章环境样品前处理技术，主要内容包括气体样品采集方法、水样采集与保存、土壤样品采集与保存和生物样品采集与保存；第四章环境化学实验，包括七个大气环境化学实验、八个水环境化学实验、五个土壤环境化学实验和五个化学物质的生态效应实验。

本书可作为高等院校环境科学、化学、生物专业本科生教材，也可作为其他相关专业本科生、研究生以及环境科学工作者的参考用书。

图书在版编目（CIP）数据

环境化学实验 = Experiments in Environmental Chemistry：英、汉 / 弓爱君主编；邱丽娜，张玮玮副主编. -- 北京：化学工业出版社，2025.1. --（普通高等教育教材）. -- ISBN 978-7-122-46827-7

Ⅰ. X13-33

中国国家版本馆 CIP 数据核字第 2024TA2851 号

责任编辑：郭宇婧　满悦芝　杨　菁　　装帧设计：张　辉
责任校对：李雨晴

出版发行：化学工业出版社
　　　　　（北京市东城区青年湖南街 13 号　邮政编码 100011）
印　　装：北京天宇星印刷厂
787mm×1092mm　1/16　印张 15¼　字数 372 千字
2025 年 6 月北京第 1 版第 1 次印刷

购书咨询：010-64518888　　　　　售后服务：010-64518899
网　　址：http://www.cip.com.cn
凡购买本书，如有缺损质量问题，本社销售中心负责调换。

定　　价：59.00元　　　　　　　　版权所有　违者必究

前　言

环境化学是一门研究有害化学物质在环境介质中的存在、化学特性、行为和效应及其控制的化学原理和方法的科学。我国环境化学学科作为化学的一个重要分支，已经成为环境科学的主流与核心组成部分。

本书编者团队致力于"环境化学"课程的教学工作，具备丰富的环境科学研究与环境化学实验实践经验，2010年获批教育部双语教学示范课程建设项目（教高函〔2010〕11号）及北京科技大学教材建设项目，2014年出版《环境化学习题集》双语教材，2016年出版《环境化学》双语教材。编者在总结环境化学学科研究成果的基础上，结合北京科技大学"十四五"重大教育教学改革方向编写此书，旨在提升相关专业的教师和学生的认识水平和实验操作技能，培养学生的爱国情怀和大国工匠精神，为实现我国的"碳达峰"和"碳中和"目标作出贡献。

《环境化学实验》双语教材主要面向高等院校环境、化学、生物、材料、冶金、能源等专业的本科生，与《环境化学》双语教材和《环境化学习题集》双语教材配套使用。

本书由弓爱君任主编，邱丽娜、张玮玮任副主编。全书共四章，包括25个实验，从基础性实验到综合性、研究性实验，内容涵盖大气环境化学、水环境化学、土壤环境化学及化学物质的生态效应实验等。其中第一章绪论由张玮玮、高歌、陈越执笔；第二章常用仪器使用方法由白玉臻、王奕文、李鹏云执笔；第三章环境样品前处理技术由邱丽娜、赵伟宇、王翔海执笔；第四章环境化学实验由弓爱君、刘杨、韩孟阳、曹龙、董艺、许鑫执笔。

限于学识、文字和英语水平，书中难免存在疏漏，希望得到读者及同仁的批评和指正。

编著者
2024年10月

Experiments in Environmental Chemistry (Bilingual)

Preface

  Environmental chemistry is a science that studies the existence, chemical properties, behaviors, and effects of harmful chemicals in environmental media, as well as the chemical principles and methods for their control. The discipline of environmental chemistry in China has become an important branch of chemistry and a mainstream and core component of environmental science.

  The team of authors of this book is dedicated to teaching environmental chemistry courses and has rich experience in environmental science research and experimental practice. In 2010, it was approved by the Ministry of Education for the construction of bilingual teaching demonstration courses (Jiao Gao Han〔2010〕No. 11) and the textbook construction project of University of Science and Technology Beijing. In 2014, the bilingual textbook *Exercises in Chemistry of the Environment was published*. In 2016, a bilingual textbook titled *Chemistry of the Environment* was published. On the basis of summarizing the research achievements of environmental chemistry and combining with the major education and teaching reform direction of the 14th Five Year Plan of University of Science and Technology Beijing, we have compiled this text book. It was intended to improve the understanding level and experimental operation skills of teachers and students in related fields, cultivate students'patriotism and national craftsmanship spirit, and contribute to achieving China's goals of "carbon peak" and "carbon neutrality".

  The bilingual textbook of Experiments in Environmental Chemistry is mainly aimed at undergraduate students majoring in environment, chemistry, biology, materials, metallurgy, energy, etc. It is used in conjunction with the bilingual textbooks of *Chemistry of the Environment* and *Exercises in Chemistry of the Environment*.

  This book was compiled with Gong Aijun as the chief editor, and Qiu Lina and Zhang Weiwei as the associate editor. The book consists of four chapters, including 25 experiments, ranging from basic experiments to comprehensive and research-oriented experiments. The content covers atmospheric environmental chemistry, water environmental chemistry, soil environmental chemistry, and ecological effects experiments of chemical substances. The first chapter of the introduction was written by Zhang Weiwei, Gao Ge, and Chen Yue; Chapter 2 the usage of common instruments was written by Bai Yuzhen, Wang Yiwen, and Li Pengyun; Chapter 3 pretreatment technology of environmental samples was written by Qiu Lina, Zhao Weiyu, and Wang Xianghai; The fourth chapter of environmental chemistry experiments was written by Gong Aijun, Liu Yang, Han Mengyang, Cao Long, Dong Yi, and Xu Xin.

  Due to limitations in knowledge, writing, and English proficiency, there are inevitably flaws in the book. We hope to receive criticism and corrections from readers and colleagues.

Editor

2024. 10

环境化学实验（双语）

目 录

第一章　绪论　　1

第一节　环境化学实验的目的和任务　　1
第二节　实验室安全守则　　1
第三节　实验室意外事故处理　　3

第二章　常用仪器使用方法　　5

第一节　玻璃仪器的洗涤及干燥方法　　5
　一、玻璃仪器的洗涤　　5
　二、玻璃仪器的干燥　　6
第二节　环境化学实验用水及制备　　6
　一、蒸馏水　　6
　二、去离子水　　7
第三节　环境化学实验常用仪器　　9
　一、天平　　9
　二、3200型原子吸收分光光度计　　11
　三、2100P型便携式浊度仪　　15
　四、FB2200型酸度计　　17

第三章　环境样品前处理技术　　19

第一节　气体样品采集方法　　19
　一、气溶胶（烟雾）样品的采集　　19
　二、室内空气污染物样品的采集　　20
第二节　水样采集与保存　　21
　一、水样的采集　　21

二、水样的预处理与保存 22

第三节　土壤样品采集与保存 25
　一、土壤样品的采集 25
　二、土壤样品的预处理与保存 25

第四节　生物样品采集与保存 26
　一、生物样品的采集 26
　二、生物样品的预处理与保存 27

第四章　环境化学实验　28

第一节　大气环境化学实验 28
　实验一　城市大气气溶胶中多环芳烃的污染分析 28
　实验二　室内空气中苯的污染分析 31
　实验三　室内空气中甲醛的浓度水平 33
　实验四　环境空气中 SO_2 液相氧化模拟 37
　实验五　空气中 SO_2 的测定 39
　实验六　盐酸萘乙二胺分光光度法测定空气中氮氧化物含量 43
　实验七　靛蓝二磺酸钠分光光度法测定环境空气中的臭氧含量 47

第二节　水环境化学实验 50
　实验八　饮用水中余氯的测定（碘量法滴定） 50
　实验九　水中 Cl^- 的测定（沉淀滴定） 52
　实验十　Fenton 试剂催化氧化染料废水 55
　实验十一　天然水中铜的存在形态 57
　实验十二　废水中生化需氧量（BOD_5）的测定 60
　实验十三　工业废水中铬的测定（二苯碳酰二肼分光光度法） 61
　实验十四　混凝实验 64
　实验十五　非色散红外吸收法测定水中总有机碳含量 67

第三节　土壤环境化学实验 70
　实验十六　石墨炉原子吸收光谱法测定土壤中的铅 70
　实验十七　重金属污染土壤的化学修复（EDTA 对土壤中铜的淋洗） 72
　实验十八　土壤有机质的测定 76
　实验十九　土壤阳离子交换量的测定 79
　实验二十　火焰原子吸收法测定工业固体废物中铜、锌、铅、镉的含量 81

第四节　化学物质的生态效应实验 84
　实验二十一　底泥对苯酚的吸附作用 84
　实验二十二　水中有机物的挥发速率 87
　实验二十三　底泥中铬的简单状态鉴别 89
　实验二十四　沉积物中重金属的存在形式和迁移规律的研究 92
　实验二十五　氢化原子荧光光度法测定食品中总砷的含量 95

参考文献　98

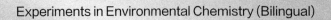
Experiments in Environmental Chemistry (Bilingual)

Contents

Chapter 1 Introduction 99

Section 1 Purpose and task of environmental chemistry experiments 99
Section 2 Laboratory safety rules 100
Section 3 Handling of laboratory accidents 102

Chapter 2 The usage of common instruments 104

Section 1 Washing and drying methods of glass instruments 104
  1 Washing of glass instruments 104
  2 Drying of glass instruments 105
Section 2 Water for environmental chemistry experiments and its preparation 106
  1 Distilled water 106
  2 Deionized water 106
Section 3 Common instruments for environmental chemistry experiments 109
  1 Balance 109
  2 3200 Atomic absorption spectrophotometer 113
  3 2100P Portable turbidimeter 118
  4 FB2200 Acidometer 121

Chapter 3 Pretreatment technology of environmental samples 123

Section 1 Collection methods of gas samples 123
  1 Collection of aerosol (smog) samples 123
  2 Collection of indoor air pollutants samples 125
Section 2 Collection and preservation of water samples 126

1	Collection of water samples	126
2	Pretreatment and preservation of water samples	128

Section 3　Collection and preservation of soil samples　132
   1　Collection of soil samples　132
   2　Pretreatment and preservation of soil samples　133

Section 4　Collection and preservation of biological samples　134
   1　Collection of biological samples　134
   2　Pretreatment and preservation of biological samples　135

Chapter 4　Environmental chemistry experiments　137

Section 1　Atmospheric environmental chemistry experiments　137
   Experiment 1　Pollution analysis of PAHs in urban atmospheric aerosol　137
   Experiment 2　Pollution analysis of benzene in indoor air　141
   Experiment 3　Concentration level of formaldehyde in indoor air　145
   Experiment 4　Liquid phase oxidation simulation of SO_2 in ambient air　150
   Experiment 5　Determination of SO_2 in air　152
   Experiment 6　Determination of nitrogen oxides in air with N-(1-naphthyl) ethylenediamine dihydrochloride spectrophotometry　158
   Experiment 7　Determination of ozone in ambient air using sodium indigo disulfonate spectrophotometry　164

Section 2　Water environmental chemistry experiments　168
   Experiment 8　Determination of residual chlorine in drinking water (iodimetric titration)　168
   Experiment 9　Determination of Cl^- in water (precipitation titration)　171
   Experiment 10　Fenton reagent catalyzed oxidation of dye wastewater　174
   Experiment 11　Existing forms of copper in natural water　177
   Experiment 12　Determination of biochemical oxygen demand (BOD_5) in wastewater　180
   Experiment 13　Determination of chromium in industrial wastewater (diphenylcarbazide spectrophotometry)　182
   Experiment 14　Coagulation experiment　186
   Experiment 15　Determination of total organic carbon in water by nondispersive infrared absorption method　190

Section 3　Soil environmental chemistry experiments　195
   Experiment 16　Determination of lead in soil by graphite furnace atomic absorption spectrometry　195
   Experiment 17　Chemical remediation of soil contaminated by heavy metals (EDTA leaching of copper from soil)　197
   Experiment 18　Determination of soil organic matter　203
   Experiment 19　Determination of soil cation exchange capacity　206

Experiment 20　Determination of copper, zinc, lead, and cadmium in industrial solid waste by flame atomic absorption spectrometry ... 210

Section 4　Ecological effects experiments of chemical substances ... 214

Experiment 21　Adsorption of phenol by sediment ... 214

Experiment 22　Evaporation rate of organic matter in water ... 218

Experiment 23　Simple identification of chromium in sediment ... 220

Experiment 24　Study on the existence forms and migration rules of heavy metals in sediment ... 224

Experiment 25　Determination of total arsenic content in food by hydride atomic fluorescence spectrometry ... 229

References ... 233

第一章
绪 论

针对日益严峻的环境污染和资源短缺问题,可持续发展、环境保护已成为时代的主题。对环境保护的迫切要求,导致环境科学类人才的需求量大幅增长,考核标准和质量要求也随之提高,新时期环境类专业人才的培养也面临着严峻的考验和挑战。

解决环境问题的重要方法和手段离不开环境化学方面的理论知识和相应的专业技能,所以环境化学在环境科学与工程及其相关领域有着非常重要的地位。其实验环节是培养学生在环境化学方面创新精神和实践能力的主要途径之一,在培养环境科学专业创新人才方面具有不可替代的重要作用。

第一节 环境化学实验的目的和任务

环境化学实验涵盖环境分析、环境监测以及环境化学的相关内容。以有关化学实验理论为基础,以化学实验在相关专业的应用为重点,其目的是通过实际操作和思考,培养学生解决问题的能力(包括有关知识的掌握及合理运用、操作技能和使用现代化仪器设备的能力、观察能力、科学研究与创新能力、独立处理突发事件的能力等)。它不是理论教学的简单验证过程,而是与理论教学相辅相成、互为补充的一门自成体系的课程。

环境化学实验的主要任务是对单个污染物的分析研究。在环境污染物分析领域,除常见的化学物质定量分析方法之外,各种仪器分析方法不断涌现,如分子光谱仪器分析、电化学仪器分析、色谱仪器分析、原子光谱仪器分析,还有流动注射分析、色谱-质谱联用等一些新的分析测试手段和技术。另外还有环境物质的理化性质分析、综合实验与环境监测等。

第二节 实验室安全守则

化学药品中有许多是易燃易爆、有腐蚀性和有毒的。因此,为保证安全,首先要求每个同学在思想上高度重视安全问题,实验前应充分了解相关安全知识,实验时要有条不紊,严格遵守实验室安全操作规程，以防事故发生。

（1）一切盛有药品的试剂瓶应贴有标签。剧毒药品必须制订保管、使用制度，严格遵守，并设专柜加锁保管。挥发性有机药品应放在通风良好的处所、冰箱或铁柜内。爆炸性药品，如高氯酸、高氯酸盐、过氧化氢以及高压气体等，应放在阴凉处保管，不得与其他易燃物放在一起，移动或启用时不得剧烈震动。高压气瓶的减压阀严禁油脂污染。

（2）严禁将食物带入实验室，使用实验器皿作为餐具，试剂入口（无论有毒与否），以及在实验室内饮食、抽烟。有毒试剂不得接触皮肤和伤口，更不能进入口内。用移液管吸取有毒样品（如铝盐、钡盐、铅盐、砷化物、氰化物、汞及汞的化合物等）及腐蚀性药品（如强酸、强碱、浓氨水、浓过氧化氢、冰醋酸、氢氟酸和溴水等）时，应用吸耳球操作，不得用嘴。有毒废液不允许随便倒入下水管道，应回收集中处理。

（3）产生有毒、有刺激性气体（如 H_2S、Cl_2、Br_2、NO_2、CO 等）的实验以及使用 HNO_3、HCl、$HClO_4$、H_2SO_4 等浓酸或使用汞、磷、砷化物等毒物时，应在通风橱内进行。当需要嗅气体的气味时，严禁用鼻子直接对着瓶口或试管口，应当用手在瓶口或管口轻轻扇动，使气体从侧面吹向自己，并保持适当距离进行嗅。

（4）开启易挥发的试剂瓶时（尤其夏季），不可将瓶口对着自己或他人的脸部。由于开启瓶口时会有大量气体冲出，如果不小心容易引起伤害事故。

（5）使用浓酸、浓碱、溴、洗液等具有强腐蚀性试剂时，切勿溅在皮肤和衣服上，必要时应戴上防护眼镜和橡胶手套。稀释浓硫酸时，必须在耐热容器内进行，应将浓硫酸缓慢倒入水中，切记不能将水倒入浓硫酸，以免迸溅。溶解 NaOH、KOH 等时会放热，也必须在耐热容器内进行。如需要将浓酸和浓碱中和时，必须先行稀释。

（6）使用易燃的有机试剂（如乙醇、丙酮等）时，必须远离火源，用完立即盖紧瓶塞。钾、钠、白磷等在空气中易燃烧的物质，应隔绝空气存放（钾、钠保存在煤油中，白磷保存在水中），取用时必须用镊子夹取。

（7）加热和浓缩液体的操作应十分小心，不能俯视正在加热的液体，更不能将正在加热的试管口对着自己或别人，以免液体溅出伤人。浓缩溶液时，特别是有晶体出现之后，要不停地搅拌，避免液体迸溅入眼睛或溅在皮肤和衣服上。

（8）实验中如需加热易燃药品或用加热的方法排除易燃组分时，应在水浴锅中或电热板上缓慢进行，严禁用电炉或明火直接加热。

（9）腐蚀性物品严禁在烘箱内烘烤。

（10）加热试管应使用试管夹，不允许手持试管加热。加热至红热的玻璃器件（玻璃棒、玻璃管、烧杯等）不能直接放在实验台上，必须放在石棉网上冷却。由于灼热的玻璃与冷玻璃在外表上没有什么区别，因此特别注意不要错握热玻璃端，以免烫伤。

（11）对于性质不明的化学试剂，严禁随意混合。严禁氧化剂与可燃物一起研磨，严禁在纸上称量 Na_2O_2 或性质不明的试剂，以免发生意外事故。

（12）玻璃管（棒）的切割，玻璃仪器的安装或拆卸，塞子钻孔等操作，往往容易割破手指或弄伤手掌，应完全按照安全使用玻璃仪器的有关操作规程进行相关操作。玻璃管或玻璃棒在切割后应立即烧圆，往玻璃管上安装橡胶管时，应先用水或甘油浸润玻璃管，再套橡胶管。玻璃碎片要及时清理，以防止事故的发生。

（13）实验室所有药品不得携带至实验室外。

（14）实验完毕后，应关闭水、电、煤气，整理好实验用品，把手洗净，方可离开实验室。

第三节 实验室意外事故处理

实验中一旦发生意外事故，应积极采取以下措施进行救护。

（一）酸烧伤

若皮肤沾上少量酸液，应用大量水冲洗。如果烧伤较重，水冲洗之后应用饱和 $NaHCO_3$ 溶液冲洗，然后再用水冲洗并涂抹凡士林油膏。若酸液溅入眼内，立即用大量水冲洗，冲洗时水流不要直射眼球，也不要揉搓眼睛，冲洗后再用 $2\% Na_2B_4O_7$ 溶液或 $3\% NaHCO_3$ 溶液洗眼，最后用蒸馏水冲洗。烧伤严重者，临时处理后应立即送往医院救治。

（二）碱烧伤

若皮肤沾上碱液，可用大量清水冲洗，直至无滑腻感，也可用稀 HAc、2% 硼酸溶液冲洗伤处之后，再用水冲净，并涂敷硼酸软膏。若碱液溅入眼内，立即用大量水冲洗，再用 $3\% H_3BO_3$ 溶液淋洗，最后用蒸馏水冲洗。

（三）溴烧伤

若遇溴烧伤，可用乙醇或 $10\% Na_2S_2O_3$ 溶液洗涤伤口，再用水冲洗干净，并涂敷甘油。

（四）磷灼伤

用 $5\% CuSO_4$ 溶液洗涤伤口，并浸过 $CuSO_4$ 溶液的绷带包扎，或用 1∶1000 的 $KMnO_4$ 湿敷，外涂保护剂并包扎。

（五）吸入刺激性或有毒气体

若吸入 Cl_2、Br_2、HCl 等气体，可吸入少量酒精和乙醚的混合蒸气以解毒。吸入 H_2S 气体感到不适或头晕时，应立即到室外呼吸新鲜空气。

（六）误食毒物

误食毒物，必须催吐、洗胃，再服用解毒剂。催吐时可喝少量（一般 15~25mL，最多不超过 50mL）$1\% CuSO_4$ 或 $ZnSO_4$ 溶液，内服后，用手指伸入咽喉部，促使呕吐，吐出毒物，然后立即送医院治疗。

（七）热烫伤

烫伤后，可先用冷水冲洗降温或用药棉浸润浓（90%~95%）酒精溶液轻涂伤处，也可用高锰酸钾或苦味酸溶液揩洗灼伤处，然后涂上烫伤膏、万花油或凡士林油。如起水泡，不要弄破，防止感染。烫伤严重的应立即送医院治疗。

（八）割伤

被玻璃割伤时，伤口内若有玻璃碎片，须先挑出，然后用消毒棉棒清洗伤口或用碘酒消毒，撒上消炎粉或敷上消炎膏，并用创可贴或绷带包扎。若伤口大量出血，应在伤口上部包扎止血带止血，避免流血过多，并立即送医院救治。

（九）触电

遇有触电事故，应立即切断电源或用木棍等绝缘物体将电源线拨开，触电者脱离电源后，必要时可进行人工呼吸。

（十）起火

如遇起火应立即灭火，同时移走火源附近的易燃药品，并切断电源，采取一切可能的措施防止火势的蔓延。一般小火可用湿布、防火布或沙土覆盖燃烧物灭火。火势较大时，可根据起火原因选择适当的灭火器材进行灭火。如四氯化碳灭火器，适用于电器失火，但是禁止用于扑灭 CS_2 的燃烧，否则会产生光气一类的有毒气体（CS_2 的燃烧可用水、二氧化碳或泡沫灭火器扑灭）；干粉灭火器，适用于扑救油类、可燃气体、电器设备、精密仪器、文件记录和遇水燃烧等物品的初期火灾；二氧化碳灭火器，适用于电器灭火；泡沫灭火器，适用于油类着火，但在电线或电器着火时禁用。

注意：油类、电线、电器设备、精密仪器等着火时，严禁用水灭火，以防触电，油可能随水漂流，扩大燃烧面积。当身上衣服着火时，应立即脱下衣服或就地卧倒打滚，或用防火布覆盖着火处。扑救蒸气有毒的化学品引起的火灾时，要特别注意防毒。

（十一）汞洒落

水银温度计打破致使汞滴落或其他原因不慎使汞洒落时，应立即用蘸水或凡士林的毛刷将汞滴集中到一起，再用吸管或拾汞棒将微小的汞滴吸起，然后在洒落汞的实验台面或地面撒硫黄粉并用力压磨（让其生成硫化汞），覆盖一段时间后再清扫。

第二章
常用仪器使用方法

第一节 玻璃仪器的洗涤及干燥方法

一、玻璃仪器的洗涤

化学实验所用的玻璃仪器是否"干净",往往会影响实验结果。此处"干净"具有纯净的含义。应重视仪器的洗涤工作。

洗涤仪器的方法很多,应根据实验的要求、污物的性质和沾污的程度来选择。一般附着在仪器上的污物包括可溶性物质、尘土和其他不溶性物质、油污和有机物,可分别采用下列洗涤方法:

（一）用水刷洗

仪器上只沾有尘土和可溶性的物质以及没有沾得很牢的不溶性物质、油污和有机物,可用毛刷直接就水刷洗。

（二）用去污粉、肥皂或合成洗涤剂洗涤

可用于洗涤沾有不溶性污物、油污和有机物的无精确刻度的仪器,如烧杯、锥形瓶、量筒等。洗涤方法是先将要洗的仪器用水润湿（水不能多）,撒入少许去污粉或滴入少量洗涤剂,然后用毛刷来回刷洗,待仪器的内外壁都经过仔细刷洗后,用自来水冲去仪器内外的去污粉或洗涤剂,要冲洗到没有细微的白色颗粒状粉末或没有洗涤剂的泡沫为止,最后,用少量蒸馏水润洗仪器三次以上,把由自来水带入的钙、镁、氯等离子洗去。注意:根据少量多次的洗涤原则,每次的蒸馏水用量都不应过多。

（三）用铬酸洗液洗涤

洗液是重铬酸钾在浓硫酸中的饱和溶液（50g重铬酸钾加入至1L浓硫酸中加热溶解得到）,具有很强的氧化性,对有机物和油污有很强的去污能力,适用于有油污的精确测量仪器或口小管细的仪器，如容量瓶、移液管、滴定管等。洗涤时先向仪器内加入少量洗液，然后边倾斜边缓慢转动仪器，使仪器内壁全部被洗液润湿，转几圈后，将洗液倒回原瓶中，然后用自来水把仪器壁上残留的洗液洗去，洗至无铬酸的黄色为止，最后用少量蒸馏水或去离子水润洗三次以上。

如果用洗液把仪器浸泡一段时间，或者用热的洗液将提高洗涤效率，但要注意安全，因为洗液有很强的腐蚀性，不要让洗液灼伤皮肤。洗液的吸水性较强，使用后应随手把装洗液的瓶子盖好，以防吸水，降低去污能力。当洗液使用到出现绿色时（重铬酸钾被还原到硫酸铬的颜色），就失去了去污能力，不能再继续使用。

注意：优先使用1和2洗涤方法清洗仪器，如非必要不得使用铬酸洗液，因为铬酸洗液价格较高，且会带来严重的污染。

另外，玻璃仪器可用超声波清洗器清洗。检验玻璃仪器是否洗干净的标准是将仪器倒置，仪器透明且内壁形成一层均匀的水膜，水不聚成水滴也不成股流下。

洗净后的仪器不得再用布或纸去擦拭，否则布或纸的纤维会留在器壁上而玷污仪器。

二、玻璃仪器的干燥

实验用的仪器除要求洗净外，有些实验还要求仪器干燥，不附有水膜。干燥的方法有以下几种：

（1）晾干。不急用的仪器在洗净后可放在仪器柜内或仪器架上，使其自然晾干。

（2）吹干。用电吹风热风直接吹干。

（3）烤干。能直接加热的仪器，如试管、烧杯、蒸发皿等可以直接在煤气灯或酒精灯上用小火烤干。烘烤试管时要注意：应把试管口向下，以免水珠倒流，炸裂试管；烘烤时应不断来回移动试管；烤到不见水珠后，再将管口朝上，赶尽水汽。

（4）烘干。洗净的玻璃仪器（不包括精度高的容量仪器）可以放在电烘箱内，控制在150℃左右烘干。仪器放进烘箱前应尽量把水倒净，并在烘箱的最下层放一搪瓷盘，防止容器滴下的水珠落到电热丝上，损坏烘箱。

（5）用丙酮、乙醇等有机溶剂快速干燥带有刻度的计量仪器（如移液管、量筒、容量瓶等）时，不能用加热的方法进行干燥，因为加热会影响仪器的精密度。可以用易挥发的有机溶剂（最常用的是乙醇或乙醇和丙酮1：1的混合物）润洗已洗净的玻璃仪器，仪器壁上的水和这些有机溶剂互相溶解混合后，倾出含水混合液（含水混合液可回收），存在的少量残留液短时间内可挥发完毕，后续晾干即可。

第二节　环境化学实验用水及制备

实验室中，清洗仪器、配制溶液、分析测定都需要大量水，但自来水含有各种杂质，不符合实验的要求。通常使用蒸馏或离子交换法获取纯净水。

一、蒸馏水

蒸馏水是自然界的水经过蒸馏器冷凝制得的。在实验室里，蒸馏水通过蒸馏烧瓶、冷凝管等仪器制取，具体操作是将自来水置于烧瓶中加热汽化，产生的水蒸气上升，经冷凝管冷凝后流入接收器进行收集，装置如图2-2-1所示。电热蒸馏水器也可供实验室使用。蒸馏水仍含有微量的杂质，在25℃时电导率为2.8×10^{-6} S/cm。

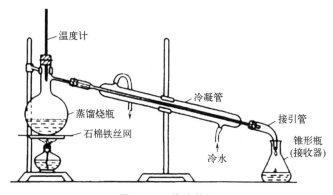

图 2-2-1 蒸馏装置

二、去离子水

用离子交换法制备的纯净水叫去离子水，目前此方法已广泛应用于各实验室和工业部门。离子交换法除用于制备去离子水外，还可用于稀有金属的分离、提纯，金属的回收，抗生素的提取等各个方面。

（一）交换原理

离子交换法是将自来水依次通过装有阳离子、阴离子交换树脂的离子交换柱，以除去水中杂质离子的一种方法。

离子交换树脂是具有网状骨架结构的固态高分子聚合物，其骨架结构上的活性官能团能与水中的离子进行交换。如聚苯乙烯磺酸型阳离子交换树脂，就是苯乙烯和一定量的二乙烯苯的共聚物，经浓硫酸处理在共聚物的苯环上引入磺酸基（—SO_3H）而成，它是强酸性的阳离子交换树脂。当树脂经过浸水，充分膨胀之后，骨架内的空隙扩大了，处在苯环上的磺酸基（—SO_3H）的氢离子便可与水中的阳离子（如 Ca^{2+}、Na^+、Mg^{2+} 等）交换。反应如下：

$$R-SO_3^- H^+ + Na^+ \rightleftharpoons R-SO_3^- Na^+ + H^+$$
$$2R-SO_3^- H^+ + Ca^{2+} \rightleftharpoons (R-SO_3^-)_2 Ca^{2+} + 2H^+$$

式中，R 是苯乙烯和二乙烯苯的共聚物。

阴离子交换树脂是指在共聚物的网状骨架上引入氨基等碱性基团，如季铵盐型强碱性阴离子交换树脂 $R \equiv N^+ OH^-$，其中 OH^- 可以与水中的阴离子 X^- 进行交换。反应如下：

$$R \equiv N^+ OH^- + X^- \rightleftharpoons R \equiv N^+ X^- + OH^-$$

（二）交换装置

通常采用离子交换柱制备纯净水，具体连接如图 2-2-2。自来水先经过阳离子交换柱除去水中的 Ca^{2+}、Mg^{2+} 等阳离子，之后流入阴离子交换柱，水中的阴离子又与交换树脂中的 OH^- 发生交换，两柱中置换出来的 H^+ 和 OH^- 结合成 H_2O。经过多次交换即可得到去离子水。

（三）制备操作

新购入的离子交换树脂中常含有低聚物、色素、灰砂等，使用时必须除去杂质，进行转型处理。

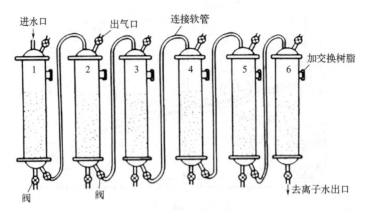

图 2-2-2　去离子水制备装置示意图
1、4—阳离子交换柱；2、3、5、6—阴离子交换柱

1. 漂洗处理

将新树脂置于盆中，用低于 40℃ 的水反复漂洗，洗至上层清液呈无色，然后将阴、阳离子交换树脂分别装入交换柱中，加水浸泡 24 小时。

2. 强酸性阳离子交换树脂的转型处理

将阳离子交换柱串联在一起，加入 2mol/L 的 HCl 将柱内的水替换出去，静置 2～3 小时，用水洗至流出液的 pH 为 3～4。再加入 2mol/L 的 NaOH 溶液，静置 2～3 小时，水洗至流出液的 pH 为 9～10。然后用 2mol/L HCl 进行转型处理，1kg 树脂约消耗酸 4.5L。

注意：转型时 HCl 的流速不可过快，以约 60mL/min 为宜。盐酸加完后，用去离子水洗至流出液的 pH 约为 4 即可。

3. 强碱性阴离子交换树脂的转型处理

将阴离子交换柱串联在一起（阴离子交换树脂的用量应是阳离子交换树脂的两倍），加入 2mol/L NaOH 溶液将柱内的水替换出去，静置 2～3 小时，用去离子水（或经过阳离子树脂的水）淋洗至流出液的 pH 为 9～10。再用 2mol/L HCl 溶液淋洗至洗出液的 pH 为 3～4。然后用 2mol/L NaOH 溶液进行转型处理，耗碱量的计算和碱液的流出速率与阳离子交换树脂的相同，当 NaOH 加完后，用去离子水淋洗至流出液的 pH 为 9～10。

将以上经活化、转型处理的离子交换树脂重新装柱，排除树脂层的气泡，即可进水作水处理，最初流出的部分水应弃去。

离子交换树脂使用一段时间后，会失去交换能力。此时应进行再生处理。处理方法为：阳离子交换柱用 HCl 溶液处理，阴离子交换柱用 NaOH 溶液处理，溶液的浓度、用量、处理方法与树脂的转型处理相同。

（四）去离子水的质量检验

酸碱度：在两支试管中加入待测去离子水，一支试管中滴加甲基红指示剂不显红色；另一支试管中滴加 0.1% 的溴百里酚蓝指示剂不显蓝色。

钙镁离子：取 10mL 待测的去离子水，加氨性缓冲溶液（将 20g NH_4Cl 加入 1L 2mol/L 的 $NH_3 \cdot H_2O$ 溶液中），调 pH 约为 10，加入铬黑 T 指示剂不显红色。

氯离子：在试管中加入待检测去离子水，用 2mol/L 的 HNO_3 酸化，再滴入 0.1mol/L $AgNO_3$ 溶液 1 滴摇匀，不得有浑浊现象。

电导率：用电导率仪测定去离子水的电导率，以判断水的质量。水的纯度越高，杂质离子的含量越少，水的电阻率越高，电导率越低。一般去离子水的电导率在25℃时为 $8.0 \times 10^{-7} \sim 4.0 \times 10^{-5}$ S/cm。

第三节　环境化学实验常用仪器

一、天平

（一）各种天平称量的准确度

实验室中使用不同等级的电子天平，各天平称量的准确度也不同。

万分之一天平，一般能称准至 0.1mg，即 0.1mg 位是不准确数。

千分之一天平，一般能称准至 1mg，即 1mg 位是不准确数。

百分之一天平，一般能称准至 10mg，即 10mg 位是不准确数。

十分之一天平，一般能称准至 0.1g，即 0.1g 位是不准确数。

（二）电子天平

最新一代的天平是电子天平。它利用电子装置完成电磁力补偿的调节，使物体在重力场中实现力的平衡，或通过电磁力矩的调节，使物体在重力场中实现力矩的平衡。常见电子天平的结构都是机电结合式的，由载荷接受与传递装置、测量与补偿装置等部件组成。可分成顶部承载式和底部承载式两类，目前常见的大多数是顶部承载式的上皿天平。从天平的校准方法来分，有内校式和外校式两种，前者是标准砝码预装在天平内，启动校准键后，可自动加码进行校准，后者则需人工去拿标准砝码放到秤盘上进行校正。梅特勒-托利多 L-IC 系列天平的构造如图 2-3-1。

（三）天平的校准

为了获得准确的称量结果，必须进行校准以适应当地的重力加速度。以下情况校准是必要的：首次使用天平称量之前；称量工作中定期进行；改变天平放置位置后。天平校准有两种校准方式：

1. 内部校准

让秤盘空着，按住"CAL"键不放，直到在显示屏上出现"CAL Int"字样后松开该键，天平自动进行校准。当在显示屏上短时间出现（闪现）信息"CAL Done"，紧接着出现"0.00g"时，天平校准结束。天平又回到称量工作方式，等待称量。内部校准见图 2-3-2。（注：内校时请确认 MENU 中 CAL 项为 CAL Int。）

2. 外部校准

准备好校准用校准砝码。让秤盘空着，按住"CAL"键不放，直到在显示屏上出现"CAL"字样后松开该键。所需的校准砝码值会在显示屏上闪烁。放上校准砝码（秤盘的中心位置），天平自动进行校准。当

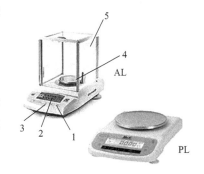
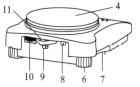

图 2-3-1　天平的构造

1—操作键；2—显示屏；3—具有以下参数的型号标牌："Max"表示最大称量，"d"表示实际分度值；4—秤盘；5—防风罩；6—水平调节螺丝；7—用于下挂称量的秤钩孔；8—交流电源适配器插口；9—RS232c 接口；10—防盗锁链接环；11—水平泡

"0.00g"闪烁时，移去砝码。当在显示屏上短时间出现（闪现）信息"CAL Done"，紧接着又出现"0.00g"时，天平的校准结束。天平又回到称量工作方式，等待称量外部校准见图 2-3-3。（注：外校时请确认 MENU 中 CAL 项为 CAL。）

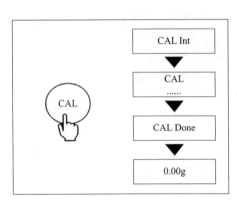

图 2-3-2　天平的内部校准

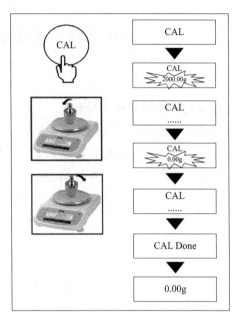

图 2-3-3　天平的外部校准

（四）称量

1. 开机、关机

开机：在开机前要查看水平仪，如不水平，请教师通过水平调节螺丝调至水平。接通电源，预热 20min（若天平长时间未使用）后方可开启显示器进行操作使用。

让秤盘空载并点击"ON"键，天平进行显示自检（显示屏上的所有字段短时间点亮）。当天平回零时，可进行称量。

关机：按住"OFF"键不放直到显示屏上出现"OFF"字样，再松开键。具体见图 2-3-4。

2. 简单称量

将称量样品放在秤盘上，等待，直到稳定状态探测符消失。读取称量结果。具体见图 2-3-5。

3. 快速称量（降低读数精度）

天平允许降低读数精度（小数点后的位数）以加快称量过程。天平默认在正常精度和正常速度状态下工作。按"1/10d"键，天平在较低的读数精度状态下工作（小数点后少一位），但是能更快地显示出结果。再点击一下"1/10d"键，天平又返回到正常读数精度工作状态。具体见图 2-3-6。

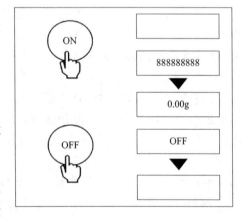

图 2-3-4　天平的开关机

4. 去皮

将空容器放在天平的秤盘上，显示该质量。点击"→O/T←"键，给容器装满称量样品，天平则显示净重。如果将容器从天平上拿走，则皮重以负值显示。皮重将一直保留到再次按"→O/T←"键或天平关机为止。具体见图2-3-7。

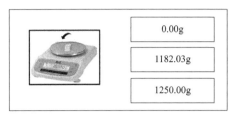

图 2-3-5 天平的简单称量

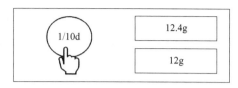

图 2-3-6 天平的快速称量

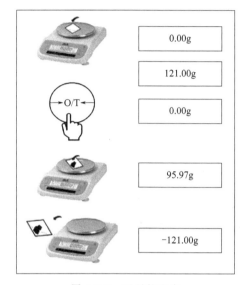

图 2-3-7 天平的去皮

5. 注意事项

（1）在天平开机前，先检查天平的秤盘及内部是否干净，如存在灰尘等应先用毛刷进行清理。

（2）天平的开机、通电预热、校准均由专人负责完成，使用人称量时只须按"ON"键、"CAL"键及"OFF"键即可，不允许乱按其他键。

（3）天平底座较轻，稳定性不足，容易被撞移位，造成不水平，从而影响称量结果。在使用时要特别注意，动作要轻、缓，并要经常查看水平仪。

二、3200型原子吸收分光光度计

（一）原理

原子吸收分光光度法是利用物质产生的原子蒸气对特定谱线的吸收作用进行定量分析的一种方法。和分光光度法一样，根据朗伯-比尔定律，吸光度为：

$$A = \lg \frac{I_0}{I} = K'LN \tag{2-3-1}$$

式中　A——吸光度；

　　　I_0——入射光强度；

I——透过光的强度;

K'——常数;

L——原子蒸气的厚度,cm;

N——试样中基态原子的数目。

原子吸收火焰温度小于 3000 K,此时火焰中极大部分是基态原子,激发态很少。基态原子数目与被测元素的浓度成正比,且火焰宽度(L)是一定的。因此,吸光度与被测元素的浓度的关系可表示为:

$$A = KC \tag{2-3-2}$$

式中　A——吸光度;

　　　K——常数;

　　　C——被测元素的浓度,mol/L。

式(2-3-2)是定量分析的基本公式。

原子吸收分光光度计一般装置主要包括:光源、原子化器、分光系统和检测显示系统。

1. 光源

目前主要使用空心阴极灯,以提供所使用的锐线光源。阴极由被测元素金属制成空心状,低压密封在一个圆柱形玻璃管内。玻璃管内充2mm汞柱的惰性气体,前端是石英窗(图 2-3-8)。

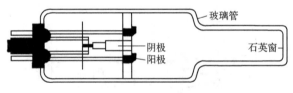

图 2-3-8　空心阴极灯

2. 原子化器

原子化器分为火焰原子化器和非火焰原子化器。火焰原子化器由喷雾器、雾化室和燃烧器组成,用来使待测样品中被测元素转变为不化合、不激发、不电离和不缔合的基态原子。非火焰原子化器应用最广泛的是石墨炉原子化器,由石墨管(杯)、炉体(保护气系统)、电源组成,利用电热、阴极溅射、高频感应或激光等方法使试样中待测定元素原子化。其灵敏度比火焰法高几个数量级,绝对灵敏度可达 10~12g,试样用量少,一般只需 1~100µL,但测定精密度较差。

3. 分光系统

原子吸收光谱仍存在某些干扰,尤其是多谱线元素更为显著,因此分光装置将共振吸收线与其他有干扰的非吸收线分离,目前常采用光栅分光。

4. 检测显示系统

由光电倍增管、放大器组成。先把光信号变为电信号,再将被测信号放大,同时去掉非被测信号,最后在数字显示窗或记录仪上显示。

(二)分析方法

1. 标准曲线法

配制一系列不同浓度的标准溶液,在所选定条件下,先用空白溶液调节吸光度为零,然后测量各标准溶液的吸光度,绘制吸光度-浓度的标准曲线。试样也须在相同条件下进行测量，将结果代入标准曲线查得浓度。

2. 加入法

先测量出样品溶液的吸光度，然后在该样品溶液中加入一定量的与样品浓度相近的标准溶液，再测出此时的吸光度。于是：

$$A_x = KC_x \tag{2-3-3}$$

$$A_0 = K(C_0 + C_x) \tag{2-3-4}$$

即

$$C_x = C_0 \times \frac{A_x}{A_0 - A_x} \tag{2-3-5}$$

式中　A_x——被测溶液吸光度；
　　　A_0——标准溶液的吸光度；
　　　C_x——被测溶液浓度，mol/L；
　　　C_0——标准溶液浓度，mol/L；
　　　K——常数。

由此可算出样品溶液的浓度，这种方法可消除基体效应。

3. 外推法

取数份体积相同的样品溶液，从第二份开始加入不同比例的标准溶液，然后稀释至一定体积，则各溶液浓度将分别为 C_x，$C_x + C_0$，$C_x + 2C_0$，$C_x + 3C_0$，$C_x + 4C_0$，测得相应吸光度分别为 A_x，A_1，A_2，A_3 和 A_4。将 A 对 C 作图，外推直线与浓度轴的交点即原来试样中某元素的浓度，见图 2-3-9。

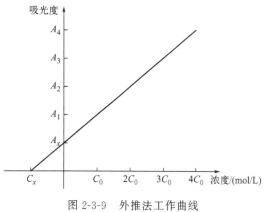

图 2-3-9　外推法工作曲线

（三）操作程序

1. 仪器操作步骤

（1）仪器在通电前应严格检查电路外接线，并将电子控制面板上所有按键开关松开，见图 2-3-10。

（2）打开仪器左上方门，把待测元素空心阴极灯插入灯座，放到灯架上用弹簧压住，使阴极灯与灯架上标记大致相平。

（3）如图 2-3-10，按下主机电源开关（12）、灯电源开关（39），待空心阴极灯发光后，调节灯电流调节旋钮（37），使空心阴极灯电流位于选定值。

（4）狭缝钮放在第二档，如图 2-3-10，选择波长扫描按（13、14）钮和波长变换按钮（17），从波长显示窗（18）上观察元素至其谱线附近，松开扫描按钮，将变换按钮推入中间空位，用手缓慢拨动波长手动轮（16）在该元素最灵敏位置附近来回移动，同时观察能量表（21）情况。在拨动波长手动轮过程中，一般会出现如下两种情况。

① 能量表无反应（或在白区），调节增益旋钮（10），再来回移动波长手动轮，使能量表指示在蓝区。

② 能量表超出蓝区，调节增益旋钮（降低负高压），使能量表指示回到蓝区后，移动手动轮使能量最大，并指示在蓝区。

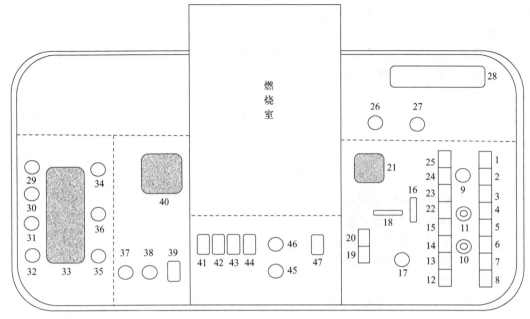

图 2-3-10　3200 型原子吸收分光光度计面板图

1—1 秒按钮；2—3 秒按钮；3—8 秒按钮；4—15 秒按钮；5—背景校正钮；6—曲线校直钮；7—火焰发射按钮；8—条件设定；9—校直钮；10—增益按钮；11—量程扩展；12—主机电源开关；13—波长扫描按钮（↓）；14—波长扫描按钮（↑）；15—参比能量按钮；16—波长手动轮；17—波长变换按钮；18—波长显示窗；19—读数按钮；20—调零按钮；21—能量表；22—浓度直接按钮；23—自动增益按钮；24—峰值保持按钮；25—积分按钮；26—光谱狭缝旋钮；27—阻尼旋钮；28—数字显示窗；29—压力表；30—雾化器旋钮；31—助燃气旋钮；32—助燃气稳压阀旋钮；33—转子流量计；34—乙炔表；35—乙炔稳压阀旋钮；36—乙炔气调节旋钮；37—灯电流调节旋钮；38—拉柄；39—灯电源开关；40—电流表；41—气源开关；42—空气-笑气转换开关；43—助燃气体开关；44—乙炔气体开关；45—燃烧器前后调节钮；46—燃烧器高低调节钮；47—手动点火开关

注：移动波长手动轮时，波长读数以长波到短波为准。自动增益按钮（23）按下时，波长无法进行调整，严重时会损坏元件。

(5) 转动空心阴极灯位置，使能量表指示最大（在无特殊情况下不要轻易调节元素灯架上螺纹套和调节螺钉）。

(6) 按下调零按钮（20），数字显示自动回零。接通记录仪电源开关，调节记录仪上零点旋钮，放下记录笔，选择走纸速度，使记录仪正常工作。

(7) 调节燃烧器前后和高低调节钮（45、46）和燃烧器灯头、手柄，使燃烧器灯头缝隙和光轴平行。校对燃烧器灯光在光轴下方适当位置（约 3～6mm）。

(8) 向预混室注水，保证废液排放管内充满水，以免由此造成漏气，并且工作过程中应一直保持废液管内有水。开启吸风罩开关，启动空气压缩机，然后打开乙炔钢瓶。

(9) 接通气路控制板上气源开关（41），调稳压阀控制空气压力为 196kPa。调节乙炔稳压阀，控制乙炔气压力为 59kPa。接通乙炔开关，调乙炔针形阀，使流量指示 1.5L/min 左右。

(10) 乙炔和助燃气通过约 5s 后，压下手动点火开关（47），点燃燃烧器上的火焰后松开。

注：熄灭火焰时，先关乙炔，后关助燃气。

（11）火焰点燃后，调整燃气流量至所需火焰状态，并预热燃烧器。

（12）喷入空白去离子水，继续预热燃烧器，此后除吸喷样品外，空白去离子水喷雾不应长时间中断。

（13）仔细选择分析条件（见仪器工作条件选择），依次完成标准试样及样品的测定。

（14）工作结束后，用去离子水喷雾喷洗，然后先切断燃烧气源，把管道中余量乙炔燃尽，火焰熄灭后切断空气压缩机电源。

2. 仪器工作条件选择

（1）灯电流的选择。固定条件：波长、狭缝宽度、燃烧器高度、空气流量、乙炔流量、试液提升量。

改变空心阴极灯电流，分别得到相应的吸光度 A，从而找出最适合的工作灯电流。

（2）燃烧器高度的选择。固定上述条件，仅改变燃烧器高度，测量元素吸光度 A 的变化，从而找出最适合的燃烧器高度。

（3）助燃气体与燃烧气体流量比的选择。固定上述条件，仅使乙炔的流量在 0.5～2.0L/min 间变化，测量元素吸光度 A 的变化，找出最佳乙炔流量和助燃气流量比。

（4）狭缝宽度的选择。固定其他条件，仅变换狭缝宽度，测量元素吸光度 A 的变化，找出狭缝的最合适位置。

三、2100P 型便携式浊度仪

（一）校准浊度仪

为保证精确度，在整个校准期间尽可能使用同一个样品池或四个匹配的样品池。通常应将样品池按匹配过程中所标注的方向标记插入。

（1）用稀释水将清洁的样品池冲洗几次，然后将稀释水或 StablCal＜0.1NTU 的标准液加入样品池至刻度线（约 15mL）。

注意：在该步骤中必须使用与准备标准液相同的稀释水。

（2）将样品池放入仪器的样品池盒中，使样品池上的方向标识与样品池盒前面的方向标识在一条线上。盖上池盖。按下"I/O"键。

注意：在按下"CAL"键前，请先选择信号平均模式选项（开或关），在校准模式下，信号平均功能不起作用。

（3）按下"CAL"键，屏幕上将显示"CAL"和"S0"图标（"0"将闪烁）。4 位显示值显示的是以前校准的 S0 标准液值。如果空白值被强制赋予为 0.0，显示的将是空白。按下"→"键，得到一个数字显示值。

（4）按下"READ"键，仪器将由 60 到 0 计数（如果信号平均功能开启，则由 67 到 0 计数），读取空白值并作为计算 20NTU 标准液测试值的校正因子。如果稀释水的浊度大于 0.5NTU，当计算校准时，屏幕上将出现"E1"字样，显示屏将自动递增到下一个标准液的测试。将样品池从样品池盒中取出。

注意：通过按下"→"键而不是读取稀释水读数的方法，可以将稀释水的浊度强制赋予零值。显示屏将显示"S0 NTU"，继续测试下一个标准液必须按下"↑"键。

（5）屏幕上将显示"S1"图标（"1"将闪烁）和"20NTU"或者以前校准的 S1 标准液值。如果该值不正确，按下"→"键直到需要编辑的数字闪烁，然后编辑该数值。使用"↑"键滚动到正确的数字。编辑后，将混合好的 20NTU StablCal 标准液或 20NTU

Foamazin 标准液加入清洁的样品池至标记线。将样品池放入仪器的样品池盒中，使样品池上的方向标识对准样品池盒前面的方向标识，盖上池盖。

（6）按下"READ"键，仪器将由 60 到 0 计数（如果信号平均功能开启，则由 67 到 0 计数），然后测试浊度并存储该值。显示屏将自动递增到下一个标准液的测试。将样品池从样品池盒中取出。

（7）屏幕上将显示"S2"图标（"2"将闪烁）和"100NTU"或者是以前校准的 S2 标准液值。如果显示值不正确，按下"→"键直到需要编辑的数字闪烁，然后编辑该数值。使用"↑"键滚动到正确的数字。编辑后，将充分混合好的 100NTU StablCal 标准液或 100NTU Foamazin 标准液加入到清洁的样品池至标记线。将样品池放入仪器的样品池盒中，使样品池上的方向标识对准样品池盒前面的方向标识，盖上池盖。

（8）按下"READ"键，仪器将由 60 到 0 计数（如果信号平均功能开启，则由 67 到 0 计数），然后测试浊度并存储该值。显示屏将自动递增到下一个标准液的测试。将样品池从样品池盒中取出。

（9）屏幕上将显示"S3"图标（"3"将闪烁）和"800NTU"或者是以前校准的 S3 标准液值。如果该值不正确，请按"→"键直到需要编辑的数字闪烁，然后编辑该数值。使用"↑"键滚动到正确的数字。编辑后，请将充分混合好的 800NTU StablCal 标准液或 800NTU Foamazin 标准液加入清洁的样品池至标记线。将样品池放入仪器的样品池盒中，使样品池上的方向标识对准样品池盒前面的方向标识，盖上池盖。

（10）按下"READ"键，仪器将由 60 到 0 计数（如果信号平均功能开启，由 67 到 0 计数），然后测试浊度并存储该值。显示屏将自动递增到下一个标准液的测试。将样品池从样品池盒中取出。

（11）按下"CAL"键确认校准值，仪器将自动返回到测试模式。

注意：按下"CAL"键完成校准系数的校准。如果在校准过程发生校准错误，则按下"CAL"键后将会出现错误信息。

校准浊度仪的注意事项如下：

如果在校准时按下"I/O"键，新的校准数据将会丢失，但旧的校准数据仍将用于测试。一旦进入校准模式，只有"READ""I/O""↑"和"→"键起作用。信号平均功能和范围选择模式必须在进入校准模式之前选择。

如果"E1"或"E2"出现在显示屏上，表明在校准过程中发生了错误。请检查标准液准备过程和校准过程；如有必要，请重新校准。按下"DIAG"键以清除错误信息（"E1"或"E2"）。如果想在未进行重新校准的情况下继续测试，请按两次"I/O"键恢复原校准值。如果显示"CAL？"，表明在校准过程发生了错误。原校准值也可能不能恢复。要么重新校准，要么使用当前的校准值。如果"CAL？"闪烁，表示仪器正在使用默认的校准值。

为查看校准值，请按下"CAL"键，然后按下"↑"键查看校准标准液的值。只要没有按下"READ"键且"CAL"不闪烁，校准值将不会更新。再次按下"CAL"键将返回到测试模式。

（二）浊度测试步骤

（1）用一个清洁的容器收集具有代表性的样品。将样品加入样品池至刻度线（约 15mL）。操作时小心拿住样品池的上部，然后盖上样品池盖。

注意：如果 5.5min 内没有按键，仪器将自动关闭，重新开启仪器请按"I/O"键。

（2）用不起毛的软布擦拭样品池，以除去水滴和手指印。

（3）滴加一小滴硅油，用油布擦拭，使硅油均匀分布在样品池整个表面。

（4）按"I/O"键，仪器将打开。测试时将仪器放在平坦稳定的板面上，不要手持仪器。

（5）将样品池放入仪器的样品池盒中，使菱形标记或方向标识对准样品池盒前面凸起的方向标识，盖上盖板。

（6）按"RANGE"键，选择手动或自动选择范围模式。当仪器处于自动选择范围模式时，显示屏将显示"AUTO RNG"。

（7）按"SIGNAL AVG"键，选择合适的信号平均模式。当仪器使用信号平均模式时，屏幕上将显示"SIG AVG"。如果样品引起噪声信号（即显示值不断变化），请使用信号平均模式。

（8）按"READ"键，屏幕上将显示"----NTU"，然后显示以 NTU 为单位的浊度数值。在灯信号关闭后请记录浊度值。

注意：仪器将默认最近一次选择的操作模式。如果前一次测试选择了自动选择范围和信号平均模式，在接下来的测试中将自动选择这些选项。

2100P 便携式浊度仪在工厂时已用 Formazin 一级标准液进行了校准，所以使用前不要求进行再次校准。哈希公司建议每 3 个月用 Formazin 进行重新校准或根据经验增加校准次数。仪器附带的 Gelex 二级标准液已标明了使用的基本范围，但经过 Formazin 校准的，必须在使用前重新确定其值。

四、FB2200 型酸度计

（一）酸度计的标定校正

（1）点击"退出"键，开机，进入自检程序；

（2）将已清洗的电极置入 pH 为 6.86 标准缓冲液中，静置 30s，按"校准"键开始校准，屏幕显示"$\sqrt{Auto}$"，完成第一个点校准；

（3）取出电极，清洗，选择 pH＝4.01 的缓冲溶液，按上述方法完成第二个点校准；

（4）校准结束后，仪表自动切换到校准结果界面。显示屏左上角显示笑脸"☺"，表示电极性能良好，可以正常使用；

（5）用去离子水清洗电极并吸干水分，将电极放入装有待测液的烧杯中，搅拌，并静置 30s。点击"读数"开始测量；

（6）仪器发出"滴"的一声且屏幕显示"$\sqrt{Auto}$"；

（7）稳定约 10s，屏幕显示数值即为最终测量值，记录数据；

（8）取出电极，清洗。

在酸度计的软件设计中实施了 6.86/4.01 和 6.86/9.18 的两组二点标定校正方案，以供在 0～14.00pH 范围内依据要求进行应用。

（二） pH 测量

在测量之前可以先回测 6.86 和 4.01 的标准液，检验仪器的测量精度是否在要求范围内。如有不符，可以重复上述标定校正步骤，对仪器实施二次标定和校正，也可以针对零点或斜率实施单点重复标定或校正（一次标定未能到位的原因可能是电极受温度因素的影响）。

在完成标定校正作业和验证步骤后，便可将已清洗待用的电极置入被测溶液中并充分搅拌，使 pH 传感器的感应球泡与被测溶液充分接触。当显示屏上的显示值稳定后，便可读取数值。

建议：从标定、校正至测量的过程尽可能在磁力搅拌器的配合下进行；如果被测溶液中含有有机物或干扰氢离子活度的因素，应先行妥当处理，避免仪器的测量结果出现不稳定的现象；为确保测量的客观性，从仪器的标定、校正到测量，电极的清洗用水务必为去离子水或蒸馏水。

仪器具有记忆当前标定校正值的功能。当频繁使用时，在不更换电极和不改变校正值的前提下，重复使用该仪器测量时，可以省去每一次的标定及校正，前提是可以关闭仪器电源，但不可以切断电源。

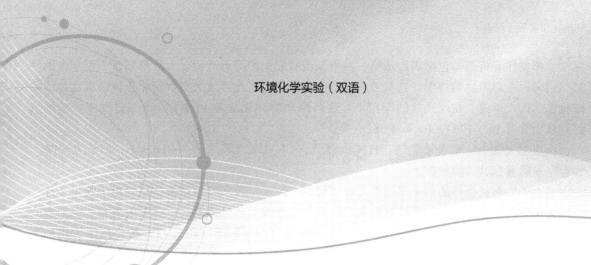

第三章
环境样品前处理技术

第一节　气体样品采集方法

一、气溶胶（烟雾）样品的采集

环境中气溶胶含有 0.01~10.0μm 的颗粒，有时甚至更大，这些气溶胶的组成，通常与颗粒大小有关。批量采样通常用过滤、碰撞或静电吸附的方法来收集气溶胶。选择方法的主要依据是避免气体颗粒在滤膜上转化，避免样品经过空气动力学作用而消失，避免由过滤材料而引起的沾污。过滤是最常用的手段，可以使用高速采样器（气体流量为 0.1~3m³/h）。滤膜的选择非常重要，需要考虑颗粒的大小、收集效率及可能的沾污等。塑料纤维滤纸机械强度较差，使用纤维滤纸和石英滤纸是较好的选择，膜过滤适用性更为普遍，这种技术具有很好的收集效率和较低的微量元素背景，但具有较高的空气流动阻力。

用滤膜采样器采集大气中的汞、有机汞和其他易挥发的元素，如铅、砷、硒等，存在一些问题，主要是滤膜表面的工作条件易导致采集样品的解吸和挥发。无论是高速或低速采样装置都面临这一问题。对于高速采样器，由于需要采样面积更大的滤纸，表面流速与低速采样器相似。

一些商用采样器通常采用碰撞收集大气颗粒，这些仪器具有按照不同大小组分采集气溶胶的能力。静电吸附在分析化学中使用较少，但具有很好的收集功能。用于职业健康调查研究中采集大气中某种颗粒的商品仪器是比较多的，它们大多采用滤膜式并可以随身携带。目前大气中微量金属研究主要集中在研究汞和铅的化学行为，对于有机铅化合物，研究它们在大气中的存在及化学行为非常重要，虽然目前从总体上讲，使用铅作为汽油防爆剂的用量已明显减少，但从冶炼厂中排放或辐射的有机铅仍然有一定的量。有机汞由于天然的和人为的排放而广泛地存在于大气中，其有害的化学性质，成为重点研究的对象。对于这两个元素，化学形态分析不仅在于研究它们在大气中的存在和反应途径，更重要的是研究其被吸入人体的毒性。这两种元素可以作为挥发性气体形态附着在颗粒物上，或者自身作为离散的颗粒物存在。挥发性有机铅，包括四乙基铅、四甲基铅和它们的降解产物，大约占城区中铅总量的 1%～4%。在含铅物燃烧中，二氯乙烯或二溴乙烯的存在会导致大多数铅化合物以氯化物或溴化物的形式排放。在硫酸盐的作用下，这些卤化物转化成为 $PbSO_4$。冶炼过程中，排放的气体中含有 Pb、PbO、PbS、$PbSO_4$ 等。

在周围大气中，元素汞蒸气、$HgCl_2$ 蒸气、CH_3HgCl 蒸气和 $(CH_3)_2Hg$ 都已发现单独存在或附着在其他颗粒物上。

对大气中有机铅样品的采集已有过许多研究。玻璃纤维和膜过滤被广泛地用于收集不挥发的无机铅盐，空气中存在的烷基铅化合物则可以通过上述过滤器。一些吸附剂和萃取剂被大胆地用于有机铅的收集和分离。氯化碘是一个有效的吸附剂，在 EDTA 的作用下，用四氯化碳萃取二烷基铅的双硫腙螯合物可以从无机铅中区分有机铅，无机铅仍留在溶液中。另外，用色谱填料来分离有机铅也是一种常用的选择。

大气中含汞样品的采集，包括不同的气体阱或吸附方法，使用不同的吸附剂可以选择性地吸附汞的多种形态。如用硅烷化的 Chromosorb W 吸附 $HgCl_2$；用 0.5mol/L NaOH 处理的 Chromosorb W 可以吸附 CH_3HgCl；涂银的玻璃珠可以吸附元素汞，而涂金的玻璃珠可选择性地吸附 $(CH_3)_2Hg$。对于含量很低的大气样品的采集，使用快速采样器按流量 5～50 m^3/min 计算，也需要几个小时的时间，这种方法的缺点是存在吸附样品解吸或挥发的可能性。而用低速采样器，按流速在 0.5～1.5 m^3/min 计算，有时需要几天的时间。

二、室内空气污染物样品的采集

室内空气污染物主要成分为甲醛、苯、甲苯、二甲苯、乙苯等容易挥发的有毒化合物。室内空气采样方法主要分为直接采样法和动力采样法两种。当室内空气的污染比较严重或是检测方法非常灵敏时，可选用直接采样。直接采样常用的工具有注射器、采样袋、采气管和真空采气瓶等。目前发展的固相微萃取的萃取纤维，也可用于现场直接采样。使用气相色谱方法测定室内空气污染物时，常用 100～500mL 注射器或采样袋，直接抽取空气样品后，旋转密闭进样口，带回实验室直接注入气相色谱进样口进行分析。采用特制的塑料袋采集室内空气时，首先应根据采集样品的特性选择塑料袋，常用的采样塑料袋有聚氯乙烯袋、聚乙烯袋和聚四氟乙烯袋。选用带有金属薄膜衬里（主要是衬铝）的袋子，对样品的稳定性有益。如对含有 CO 等混合样品，放置在聚氯乙烯袋中，只能稳定十几小时。同样的样品，放置在铝膜衬里的聚酯袋子里可以稳定 100h 而无损失。采用塑料袋采集室内空气时，可使用大体积进样器，但最好的选择是使用二联球。二联球方法适合采集 100～500mL 样品。也可用固定体积的容器采集气体样品，常见的容器有由耐压的透明玻璃或由不锈钢制成的真空采气瓶，其体积在 500～1000mL。

在污染物浓度较低或连续采集样品时，往往采用动力采样。动力采样方法是采用机械泵，迫使室内空气通过收集样品的介质，让被测物在介质中吸附或冷凝，达到富集采样的目的。吸收介质既可以是液体，也可以是填充柱或者是各种膜材料。溶液的吸收法是用一个气体吸收管，内装吸收液，其后端接有抽气装置并以一定的流速通过吸收管抽入空气样品，当空气通过吸收液时，由于发生了溶解作用或化学反应，被测组分就被留在吸收液里，取样结束后，倒出吸收液，测定其中的成分含量。填充柱吸附法使用较为广泛，常用的柱吸附材料是活性炭和硅胶。根据被测对象的不同，可以选择有效的吸附剂。如采集含汞样品，可选用 60～80 目的 Tenax 等。填充柱采样方法是目前应用最广泛的方法，与溶液吸收法比较，这种方法的特点有 4 个：①可以长时间连续采样，通过测定不同时段、区段的浓度变化情况，可以有效地反应污染变化的动态过程；②通过选择合适的填充剂，对于气体、蒸气和气溶胶都有较好采集效率；③污染物在填充柱上比较稳定，便于存放或携带；④适用于各种场合包括现场采样。常用于空气采样器的抽气泵有薄膜泵、电磁泵和刮板泵等。低温冷凝采样方法常用于挥发性气体的采集。常用的制冷剂有冰-盐水（-10℃）、干冰-乙醇（-72℃）、液氧（-183℃）、液氮（-196℃）和半导体制冷器（目前可达-50℃）等。此外，还有被动式采样方法等。

第二节　水样采集与保存

一、水样的采集

环境样品测定中采集最多的就是水样。环境水样可分为自然水（雨雪水、河流水、湖泊水、海水等）、工业废水及生活污水。自然界中的水含有复杂的多种成分，包括有机胶体、细菌及藻类，无机固体包括金属氧化物、氢氧化物、碳酸盐和黏土等，其中微量元素或有机污染物的含量往往是很低的。采集的各种水样必须有代表性。

（一）采集位点

工业污水中有毒化合物较多，而生活污水中有机质、营养盐等成分居多。采样时应尽可能考虑全部影响因素以及这些因素可能产生的变化情况。主要包括以下几个因素：①测定内容，即测定化合物的类别。采集前对于样品的用途应该有清楚的了解，假若是测定一条河中某种污染物长期的变化规律，一定要选取在固定间隔期间内可以重复选取的地点；②样品的大致浓度范围；③基体的种类及其均匀程度；④所用分析方法的特殊要求。影响水样性质的物理过程有：逸气、沉淀、悬浮物损坏、沉积物和悬浮物的扰动、分析物再分布、玷污等。影响采集水样性质的化学过程主要有：化学降解、光降解等。采样时应避免采样设备、船甲板或排污水的污染。

自然界中微量元素和有机污染物的含量与水样深度、盐度及排放源有关，只有个别有机金属化合物如甲基锗等与采集深度及盐度无关。

采集的各种水样必须具有代表性，能反应水质特征。河口和港湾检测断面布设前，应查河流流量、污染物种类、点或非点污染源、直接排污口污染物的排放类型及其他影响水质均匀程度的因素。监测断面的布设应具有代表性，能较真实地全面反映水质及污染物的空间分布和变化规律。对于使用管道或水渠排放的水样采集，首先必须考虑通过实验确定污染物分布的均匀性，应该避免从边缘、表面或地面等地方采样，因为通常这些部位的样品不具备代表性。供分析河流天然水化学成分的水样，一般在水文站测流断面中泓水面下 0.2~0.5m 采样，断面开阔时应当增加采样点。岸边采样点必须设在水流畅通处。入海口区的采样断面一般与径流扩散方向垂直布设。港湾采样断面（站位）视地形、潮汐、航道和检测对象等情况布设。在潮流复杂区域，采样断面可与岸线垂直设置。海岸开阔海区的采样点位呈纵横断面网格状布设。必要时还可根据不同的物理水文特征和采样要求在不同深度分层取样，一般可分为表层、10m 层和底层。

（二）采样要求

水样采集一般应使用专用采样器，以保证从规定的水深采集代表性水样。

（1）表面水样的采集，必须考虑将聚乙烯瓶插入水面以下避开水表面膜并戴上聚乙烯手套，表面水样可以用聚乙烯水桶采集。测定海水中金属元素或有机污染物时，必须更加小心注意采样器具的清洁问题。用船来采集水样，不管是大船还是小舟必须考虑来自船体自身的沾污。

（2）对于深水采样，目前采用的器皿大多是由聚乙烯、聚丙烯、聚四氟乙烯、有机玻璃（甲基丙烯酸甲酯）等加工而成，避免使用胶皮绳、铁丝绳等含有胶皮或金属的材料，避免铁锈或油脂的沾污。

（3）对于天然水样，大多采用定时采集的方法。为了反映水质的全貌，必须在不同的地点和时间重复取样。另外，在有多种排放源存在的情况下，采自不同的横断面或不同深度的样品都会有很大差别。自动采集装置主要用于高采样密度和长期连续不断采样。连续测定的常规参数主要包括 pH、电导率、盐度、硬度、浊度、黏度等。

（4）采集雨水和雪样时，如果是沉积物，可用大体积采样器同时收集湿的和干的沉积物，如果采集湿样，只能在下雨或下雪时采集。对于高山和极地雪的采集，必须用洁净的聚乙烯容器，操作者戴洁净手套，在逆风处采样。采样时先用塑料铲刮出一个深度 30cm 的斜坡，用大约 1000mL 的聚乙烯瓶横向采集离地面 15～30cm 的雪样，采集后立即封盖冷藏处理直到样品分析。

（三）采样频率

采样时间和频率的确定原则：以最小工作量满足反映环境信息所需的资料、能够真实地反映出环境要素的变化特征，尽量考虑采样时间的连续性、技术上的可行性和可能性。采集的频率必须足够大以反映水样随季节的变化。通常采用两周一次或一月两次。在确知一些排放源排放时间时，采样也可随此变化。

二、水样的预处理与保存

除非将采到的水样马上进行分析，否则在水样储存以前必须进行适当的预处理。预处理主要依据被测水样的不同要求而异。通常对于微量元素或有机分析，首先必须通过过滤或者离心将水样中的颗粒物质除去（如果测定颗粒物中的污染物成分，则须收集这部分样品），然后加入保护剂，最后水样盛放在没有污染的容器内，并贮存在合适的温度下，以防止有效成分的损失、降解或形态变化。

在未过滤的样品中，颗粒物和溶解于样品中的碎片之间的相互作用，有可能引起样品中重金属化学形态分布的变化。研究人员发现重金属在沉积物与水的混合物中的吸附-解吸平衡时间是很快的，一般少于 72h，最大吸附发生在 pH=7.5 左右。采样后，溶液平衡发生任何变化，颗粒物所提供的吸附部位都将为金属形态的迁移提供路径，而在某些条件下，解吸已吸附的金属是可能的。另外，一些研究表明，将未过滤的海水样品贮存在聚乙烯容器中，溶解的重金属组分如 Pb、Cu、Cd、Bi 等没有损失。

高的细菌浓度伴随着沉积物的存在同样也会导致水溶性金属形态的损失。细菌和藻类的生长包括光合成及氧化等作用将会改变水样中 CO_2 的含量因而导致 pH 值的变化，pH 值的变化往往带来沉淀、改变螯合或吸附行为，以及溶液中金属离子的氧化还原作用。

利用未经处理的膜来过滤海水样品中的含汞样品，可能造成 10%～30% 的损失。然而使用处理过的玻璃纤维过滤，汞的损失可降低至 7% 以下。

由于贮存样品中的细菌生长和繁殖的不可测性质，采样后的过滤越早越好。如果时间推迟至几个小时之后，样品最好在4℃左右保存以便抑制细菌的生长。

利用0.45μm的微孔膜可以方便地区分开溶解物和颗粒物，通过滤膜的过滤液中还可能含有0.001～0.1μm的微生物和细菌的胶粒以及小于0.001μm的溶解于水中的组分。0.45μm的滤膜可以滤除所有的浮游植物和绝大多数的细菌。连续的过滤有时可能造成滤膜的堵塞，这时一般需要更换新膜或是采用加压过滤。

利用过滤仪器，应该注意仪器与溶液相接触部分的材料，如硼硅玻璃、普通玻璃、聚四氟乙烯等，同时也要考虑过滤器的类型，如针孔还是加压。玻璃过滤器使用橡胶塞子容易造成沾污。一般选择使用硼硅玻璃的真空抽滤系统。过滤以前，过滤器材应用稀酸洗涤，通常可以在1～3mol/L盐酸中浸泡一夜。

未被处理过的过滤膜表面极易吸附水中的镉和铅，但用来过滤河水时，未发现上述元素浓度的变化。一般的滤膜使用前先用20mL 2mol/L HNO_3 洗涤，再用50～100mL蒸馏水冲洗。接收的烧杯或三角烧瓶必须用蒸馏水将酸冲洗干净，最初的10～20mL滤液去掉。对于海洋深水样的过滤，滤膜最好先用稀硝酸浸泡。

加压过滤或真空抽滤是通常使用的两种方法。加压过滤速度快，适用于过滤含有大量沉积物的河水水样，如果使用47mm直径、0.45μm膜过滤水样，速度大约在100mL/h左右，加压过滤通常使用超滤膜。

对于难以过滤的样品，离心也是一种有效的手段，但离心的过程容易引起污染。离心分离的效率跟离心的速度、时间以及颗粒的密度有关。

水样采集后应尽快分析，如放置过久，水中某些成分会发生变化，供理化分析用的水样允许存放时间为：洁净水，72h；轻度污染水，48h；污染水，12h。不能及时运输或尽快分析的水样，则应根据不同检测项目的要求，采取适宜的保存方法。常用水样保存技术见表3-2-1。

表 3-2-1　常用水样保存技术

	待测项目	容器类别	保存方法	可保存时间	建议
物理与化学分析	pH	P/G	—	—	现场直接测试
	酸度/碱度	P/G	在2～5℃暗处冷藏	24h	水样注满容器
	嗅	G	—	12h	—
	电导率	P/G	在2～5℃冷藏	24h	—
	色度	P/G	在2～5℃暗处冷藏	24h	—
	悬浮物	P/G	—	24h	单独定容采样
	浊度	P/G	—	—	现场直接测试
	臭氧	G	—	—	—
	余氯	P/G	NaOH固定	6h	最好现场分析
	二氧化碳	P/G	—	24h	水样注满容器
	溶解氧	—	现场固定并存放暗处	—	碘量法用1mL 1mol/L硫酸锰溶液与2mL 1mol/L碱性碘化钾
	油脂/油类	G	用HCl酸化至pH≤2	1周	—
	碳氢化合物/石油及衍生物	G	用HCl或H_2SO_4酸化，pH为1～2	1月	—

续表

	待测项目	容器类别	保存方法	可保存时间	建议
物理与化学分析	高锰酸钾指数	G	1~5℃暗处冷藏	24h	—
		P	−20℃冷冻	1月	—
	化学需氧量	G	用 H_2SO_4 酸化,pH≤2	24h	—
		P	−20℃冷冻	1月	—
	砷	P/G	1L 水样中加浓 HCl 2mL	2周	—
	生化需氧量	G	在 2~5℃暗处冷藏	尽快	—
	凯氏氮、氨氮	P/G	加 H_2SO_4 酸化,使 pH<2,并在 2~5℃冷藏	尽快	—
	硝酸盐氮	P/G	酸化至 pH<2,并在 2~5℃冷藏	尽快	—
	亚硝酸盐氮	P/G	在 2~5℃冷藏	尽快	—
	有机氯农药	G	在 2~5℃冷藏	1周	—
	有机磷农药	G	在 2~5℃冷藏	24h	—
	酚	P	用硫酸铜抑制生化作用,并用磷酸酸化,或用 NaOH 调节使 pH>12	24h	—
	叶绿素 a	BG	1~5℃冷藏	24h	—
	汞	P/G	—	2周	—
	镉、铅、铜、铝、锰、锌、镍、总铁、总铬	P/BG	硝酸酸化使 pH<2	1个月	—
	六价铬	P/G	用 NaOH 调节,使 pH=7~9	—	—
	钙、镁、总硬度	P/BG	过滤后酸化滤液,使 pH<2	数月	酸化时不要用 H_2SO_4
	氟化物	P	中性样品	数月	—
	氯化物	P/G	—	数月	—
	总磷	BG	用 H_2SO_4 酸化,使 pH<2	—	—
	硒	G/PG	用 NaOH 调节,使 pH>11	数月	—
	硫酸盐	P/G	在 2~5℃冷藏	1周	—
微生物与生物学分析	细菌总数/大肠菌群/沙门氏菌等	灭菌容器 G	在 2~5℃冷藏	尽快	—
	鉴定和计数:底栖类无脊椎动物	P/G	加入 70%(体积分数)乙醇或加入 40%(体积分数)中性甲醇	1年	—
	浮游植物 浮游动物	G	加 40%(体积分数)甲醛,成为 4%(体积分数)的福尔马林溶液,或 1 份体积样品加入 100 份卢戈耳溶液	1年	卢戈耳溶液:150g KI、100g I_2、10mL 乙醇配成水溶液

注:P—聚乙烯;G—玻璃;BG—硼硅玻璃。

第三节 土壤样品采集与保存

一、土壤样品的采集

(一)污染土壤样品的采集

1. 采样布点

选择一定数量能代表被调查地区的地块作为采样单元（1300～2000m²），在每个采样单元中，布设一定数量的采样点。同时选择对照采样单元布设采样点。

为减少土壤空间分布不均一性的影响，在一个采样单元内，应在不同方位上进行多点采样，并且均匀混合成为具有代表性的土壤样品。

2. 采样深度

一般监测土壤污染状况，只需取 0～15cm 或 0～20cm 表层（或耕层）土壤，使用土铲采样。

3. 采样量

由上述方法所得的土壤样品一般是多样点均量混合而成的，取土量往往较大，而一般只需要 1～2kg 即可，因此对所得混合样须反复按四分法弃取。最后留下所需土样的量，装入塑料袋或布袋内。

(二)土壤背景值样品采集

在每一个采样点均须挖掘土壤剖面进行采样。我国环境背景值研究协作组推荐，剖面规格一般为长 1.5m、宽 0.8m、深 1.0m，每个剖面采集 A、B、C 三层土样。过渡层（AB、BC）一般不采样。当地下水位较高时，挖至地下水露出时为止。现场记录实际采样深度，如 0～20cm、50～65cm、80～100cm。在各层次典型中心部位自下而上采样，切忌混淆层次、混合采样。

在山地土壤土层薄的地区，B 层发育不完整时，只采 A、C 层样。

干旱地区剖面发育不完整的土壤，采集表层（0～20cm）、中土层（50cm）和底层（100cm）附近的样品。

采样点数的确定：采样点的数目与所研究地区范围的大小、研究任务所设定的精密度等因素有关。在全国土壤背景值调查研究中，为使布点更趋合理，采样点数依据统计学原则，即在所选定的置信水平下，与所测项目测量值的标准差、要求达到的精度相关。每个采样单元采样点位数可按下式估算：

$$n = \frac{t^2 s^2}{d^2} \tag{3-3-1}$$

式中　n——每个采样单元中所设最少采样点位数；
　　　t——置信因子（当置信水平为 95% 时，t 取值 1.96）；
　　　s——样品相对标准差；
　　　d——允许偏差（若抽样精度不低于 80% 时，d 取值 0.2）。

二、土壤样品的预处理与保存

(一)土样风干

从野外采集的土壤样品运到实验室后，为避免受微生物的作用引起发霉变质，应立即将全部样品倒在塑料薄膜上或瓷盘内进行风干。当达半干状态时把土块压碎，除去石块、残根等杂物后铺成薄层，经常翻动，在阴凉处使其慢慢风干，切忌阳光直接暴晒。样品风干处应防止酸、碱等气体及灰尘的污染。

（二）磨碎与过筛

一般常根据所测组分及称样量决定样品细度。

1. 物理分析

取风干样品100~200g，放在木板上用圆木棍碾碎，经反复处理使土样全部通过2mm孔径的筛子，将土样混匀贮于广口瓶内作为土壤颗粒分析及物理性质的测定。

2. 化学分析

分析有机质、全氮项目，应取一部分已过2mm筛的土样，用玛瑙研钵继续研细，使其全部通过60号筛（0.25mm）。用原子吸收分光光度法（AAS）测Cu、Cd、Ni等重金属时，土样必须全部通过100号筛（尼龙筛），研磨过筛后的样品混匀、装瓶、贴标签、编号、储存。

（三）土样的保存

一般土壤样品需保存半年至一年，以备必要时查核之用。环境监测中用以进行质量控制的标准土样或对照土样则需长期妥善保存。储存样品应尽量避免日光、潮湿、高温和酸碱气体等影响。

将风干的土样、沉积物或标准土样等储存于洁净的玻璃或聚乙烯容器内，在常温、阴凉、干燥、避光、密封（石蜡涂封）条件下保存30个月是可行的。

第四节　生物样品采集与保存

一、生物样品的采集

生物样品涉及复杂的基体，这些基体既有固态的也有液态的，包括所有的水生或陆生动、植物。形态分析有时针对整个生物体，有时是其中的一部分器官或组分，有时则只测定排泄物。环境分析中，生物样品主要包括鱼类、果壳类、海藻类、草本植物、果实、蔬菜、叶子等动植物样品。在职业健康研究中，主要研究人体组织、头发、汗液、血液、尿样和粪便等。

生物样品的采集关键在于防止样品的沾污，对于金属元素的形态分析，应该避免使用金属刮刀、解剖刀、剪刀、镊子或针等。实验室用的硼硅玻璃器皿，聚乙烯、聚丙烯或聚四氟乙烯以及石英器具等可以替代上述金属制品。处理样品时，不要直接用手接触，而应戴塑料手套，带粉末的如滑石粉等应注意冲洗干净，因这些粉末往往会带进锌和其他金属。在实践中，金属刀片和活检用的针头不可避免地要用于样品处理，虽然这些刀具会带来一定的金属沾污，但比石英和玻璃刀要方便和有效得多。研究表明，使用不锈钢刀具处理活检组织，可能带入3ng/gMn、15ng/gCr和60ng/gNi，如果研究上述金属元素在生物体态中的浓度和形态，不锈钢刀具显然是不合适的。

研究人体血清中的微量元素，有时含量相差很大，这除了分析方法和个体差异外，在很大程度上来源于采样时的污染。有研究表明，用不锈钢针头采集血液样品时，前20mL中铁、锰、镍、钴、钼、铬的含量明显增高，铜和锌的含量没有什么变化。由于这个原因，最好用聚四氟乙烯和聚丙烯导管替代。

灰尘是生物样品中锌、铝和其他金属元素污染的主要来源。在体液等样品采集中应该避免带入灰尘。尿样也极易带入灰尘，所以采样时必须加倍注意，将其接收在带盖的、用酸冲洗过的聚乙烯瓶子里。

二、生物样品的预处理与保存

对许多生物样品，需要进行最初的预处理。这种预处理应该在采样后立即进行。例如，在分析贝壳类样品中的微量元素时，需要将贝壳外层的沉积物清洗干净，然后开壳，采集整个贝肉或个别部位。同样，水生植物如海藻等亦需要仔细清洗以除去沉淀物、寄生植物或其他类似的表面沾污。必须注意样品的代表性。微量元素往往在一些特殊的部位有更大的浓度，如植物根和叶子等，而且植物的大小与浓度也有关系，如果分析单个样品时，采样时就应注意这些特点。上述样品采集后如不马上分析应该冷藏。处理上述样品时应戴聚乙烯手套，样品存放于塑料袋或塑料容器中。对于头发样品，为了排除外来元素，可以使用一些清洗办法进行微量元素分析发样的处理，可以在 0.1% Triton-X100 中，用超声波振荡仪处理，过滤后用甲醇冲洗，然后在空气中用吹风机吹干。

对于血液样品的处理，需要根据全血、血浆及血清的不同要求进行前处理。全血短期内可以在 4℃ 贮存，冷冻效果较好，冷藏往往会导致沉淀出现。解冻时沉淀出的固体加酸可以溶解。

样品的均匀往往是处理大量样品的第二步，如果使用机械匀浆机，刀片的沾污问题需要考虑。少量样品也可以直接溶解在浓盐酸或强碱里，季铵盐的氢氧化物也经常用来溶解少量组织样品。如在 20% 的四甲基氢氧化铵中浸泡 2h，可以有效地从蛋白质和脂肪中游离出烷基铅化合物，这种提取方法适用于鱼、海藻和其他海生生物的处理。提取时在 60℃ 水浴中加热可以加快提取速度。

高压焖罐也经常用于样品处理，这种方法可以有效地防止样品处理时的沾污，从而保护微量元素。在高压溶样前，样品一般经冻干或风干后，再经高压处理。高压焖罐的温度控制要根据不同样品的需求而定，一般而言，在 160~180℃ 下处理 6h 即可得到满意的结果。

冷冻干燥是除去固体生物样品中水分的好办法，可以避免微量元素的损失或样品沾污，这个方法适宜少量样品。在通风橱中控制温度在 100℃ 左右连续加热，是一种最方便的方法，但是容易挥发的组分，像元素汞和其他一些有机金属化合物极易在此期间损失。

马弗炉在 500℃ 时可以分解大多数有机化合物，也适合含 Hg、As、Sn、Se、Pb、Ni、Cr 等样品的处理。

冻干或消解的样品比较容易保存，在萃取前或溶解前，需要重新混合均匀。

第四章
环境化学实验

第一节 大气环境化学实验

实验一 城市大气气溶胶中多环芳烃的污染分析

在城市大气中普遍存在的多环芳烃（PAHs）由于其高毒性、持久性、积聚性和流动性大等特点，被列入需要进行控制和治理的持久性有机污染物质。

PAHs 是指两个以上苯环以稠环形式相连的化合物。按其分子结构的不同可分成两大类，即联苯类和稠环类。通常通过高温燃烧合成多环芳烃及杂环化合物。大气气溶胶中的多环芳烃主要来源于含碳物质的不完全燃烧；机动车尾气污染源是可吸入颗粒物中多环芳烃的主要贡献源，其次是燃煤；扬尘中的多环芳烃主要来自煤烟尘的沉降。目前我国对空气中多环芳烃的研究比较多地集中于颗粒物中多环芳烃，而对大气气溶胶中多环芳烃的研究相对较少。本实验采用中流量采样法测定大气气溶胶中的多环芳烃，同时就 PAHS 在大气环境中的存在状态，即在大气气溶胶中的分布情况进行分析。

一、实验目的

(1) 掌握空气中 PAHs 的采集、提取、分析方法。
(2) 掌握高效液相色谱仪的测定原理及使用方法。
(3) 分析评价空气中 PAHs 的污染现状及形态。

二、实验原理

多环芳烃在空气中的主要存在形式包括气态、颗粒态（吸附在颗粒物上），而且一定条件下，两者间可互相转化。影响多环芳烃含量的因素包括自身物理化学形态、气温、其他共存污染物，如飘尘、臭氧等。

主要操作步骤分为以下几步：
(1) 大气自动监测子站采集大气气溶胶中的 PAHs；
(2) 二氯甲烷作为萃取剂，超声提取样品中的 PAHs；
(3) 高效液相色谱法测定 PAHs 的峰高或峰面积，外标法定量；
(4) 分析、评价 PAHs 污染水平及形态分布。

三、实验仪器与试剂

（一）仪器

高效液相色谱仪，带有 LC7060 型二极管阵列（PDA）检测器、超声波清洗器、大气自动监测子站、RAM-1020 型 β-射线测尘仪、K-D 浓缩器、0.45μm 滤膜、玻璃纤维过滤膜带（30mm×20m）。

（二）试剂

多环芳烃标准混合溶液（100mg/L），组成为苊（Ace）、芴（Fl）、菲（Ph）、蒽（An）、荧蒽（Fa）、芘（Py）、苯并[a]蒽（BaA）、䓛（Chry）、苯并[b]荧蒽（BbF）、苯并[k]荧蒽（BkF）、苯并[a]芘（BaP）、二苯并[a,h]蒽（DahA）、苯并[g,h,i]苝（BghiP）、茚并[1,2,3-c,d]芘（IcdP），均为 200μg/mL；Mili-Q 级纯水，超声脱气备用；甲醇（色谱纯）、硝酸（优级纯）、二氯甲烷（色谱纯）。

四、实验步骤

（一）采样点选择

在下面 5 个不同功能区根据实际条件选择采样点：
(1) 工业区，选择化工厂附近，选择 5~15 个采样点；
(2) 居民区，选择比较有代表性的小区住宅附近，选择 10~15 个采样点；
(3) 商业区，选择比较有代表性的商业街附近，选择 5~15 个采样点；
(4) 交通稠密区，选择城市主干道附近，选择 5~10 个采样点；
(5) 文化学校区，选择有代表性的学校、博物馆等场所附近，选择 5~10 个采样点。

（二）滤膜带的预处理

玻璃纤维滤膜带使用前在稀硝酸溶液中浸泡一周，再在 Mili-Q 水中浸泡一周后自然晾干，置于干燥器中干燥，直到滤膜恒重、备用。经该方法处理后的滤膜空白值接近 0。

（三）样品的采集

在各大气自动化子站上均利用 RAM-1020 型射线测尘仪分别采集大气气溶胶样品。用宽 20mm、长 20m 玻璃纤维滤膜带采集 48h 连续气溶胶样品，采样头距地面高度 9~18m。在各个功能区附近采样点采集大气气溶胶样品，每次采集样品数量根据分析要求，在 15~40 个之间。

（四）样品的预处理

将采集的滤膜带上的斑点剪成碎条放入 25mL 磨口锥形瓶中，加入一定量的二氯甲烷，用超声波超声萃取 3 次，每次 5min 左右，将萃取液通过 0.45μm 的滤膜过滤到 K-D 浓缩器中，在 65~70℃水浴上浓缩至试液体积约 0.5mL，然后用纯氮气吹干，以甲醇溶解残渣并定容至 0.5mL 待测。进行 HPLC 分析时，进样前需用 0.45μm 针头过滤器过滤后进样。

（五）色谱操作程序

PAHs 的高效液相色谱分析采用非固定荧光激发波长和荧光发射波长的高效液相色谱-荧光检测法测定大气气溶胶中多环芳烃类化合物，用这种方法对实际样品的分析灵敏度高、干扰少、方法简单、快速。多环芳烃化合物的分析波长见表 4-1-1。

表 4-1-1 多环芳烃化合物的分析波长

时间/s	激发波长 λ/cm	发射波长 λ/cm	测定组分
0	291	356	苊(Ace)、芴(Fl)
210	250	400	菲(Ph)、蒽(An)
305	289	462	荧蒽(Fa)
355	320	380	芘(Py)
540	266	403	苯并[a]蒽(BaA)、䓛(Chry)、苯并[b]荧蒽(BbF)、苯并[k]荧蒽(BkF)、苯并[a]芘(BaP)
900	294	430	二苯并[a,h]蒽(DahA)、苯并[g,h,i]苝(BghiP)
1530	294	482	茚并[1,2,3-c,d]芘(IcdP)

色谱柱为 PECHROMSEP，250mm×4.6mm。流动相流速为 1.0mL/min。流动相约进样 2L。流动相为甲醇-水，梯度洗脱，甲醇-水的最佳体积配比为甲醇（15%）：水（100%）=18:82，在该配比下保持 7min，然后以线性梯度洗脱 5min 后，将该体积配比改变为甲醇（15%）：水（100%）=5:95，保持 17min。

标准曲线的绘制：峰高为纵坐标，PAHs 浓度为横坐标，绘制每一种多环芳烃的标准曲线，其浓度范围应根据 HPLC 的灵敏度及样品的浓度而定。

五、数据处理

计算大气气溶胶中 PAHs 的浓度：根据色谱图，保留时间定性，峰高和峰面积定量，积分计算出浓度。测定 PAHs 流动相线性梯度如表 4-1-2 所示，可根据实际样品浓度加以选择。

表 4-1-2 PAHs 流动相线性梯度

时间/min	甲醇体积分数/%	水体积分数/%
0.0	50	50
5.5	70	30
16.0	80	20
20.0	85	15
25.0	90	10
30.0	95	5
35.0	95	5
40.0	100	0
45.0	100	0
50.0	50	50

六、思考题

(1) 分析实验数据,说明空气中PAHs的主要来源。
(2) 试述影响空气中PAHs存在形态的主要因素。
(3) 举例说明几种测定PAHs的方法及其区别。

实验二 室内空气中苯的污染分析

一、实验目的

(1) 学习气相色谱的基本操作方法。
(2) 学习建立色谱法测定室内空气中苯的方法。
(3) 了解室内通风条件对室内空气中苯含量的影响。

二、实验原理

苯是比较常见的一类污染物。室内空气中的苯通常来自室内用品中溶剂的挥发和有机物燃烧。《室内空气质量标准》规定室内空气中苯的限值是 $0.03mg/m^3$。

苯通常是无色的,有芳香气味,容易挥发。它们微溶于水,易溶于乙醇、乙醚、氯仿和二硫化碳等有机溶剂。测定室内空气中苯的浓度,可采用活性炭吸附取样或低温冷凝取样,然后用气相色谱法测定。室内空气中苯系物各种测定方法及特点见表4-1-3,本实验采用 DNP-Bentane 柱(CS_2 解吸)法。

表 4-1-3 室内空气中苯系物各种测定方法及特点

测定方法	原理	测定范围	特点
DNP+Bentane (CS_2 解吸)法	用活性炭吸附采样管富集空气中苯、甲苯、乙苯、二甲苯后,加二硫化碳解吸,经 DNP+Bentane 色谱柱分离,用火焰离子化检测器测定,以保留时间定性,峰高(或峰面积)外标法定量	当采峰面积为100L时,最低检出浓度:苯 $0.006mg/m^3$、甲苯 $0.004mg/m^3$、二甲苯及乙苯均为 $0.010mg/m^3$	可同时分离测定空气中丙酮、苯乙烯、乙酸乙酯、乙酸丁酯、乙酸戊酯,测定面广
PEG-6000柱(CS_2 解吸进样)法	用活性炭管采集空气中苯、甲苯、二甲苯,用二硫化碳解吸进样,经 PEG-6000 柱分离后,用氢焰离子化检测器检测,以保留时间定性,峰高定量	苯、甲苯、二甲苯的检出限分别为 0.5×10^{-3}、1×10^{-3}、$2\times10^{-3}\mu g$(进样 $1\mu L$ 液体样品)	只能测苯、甲苯、二甲苯、乙烯
PEG-6000柱(热解吸进样)法	用活性炭管采集空气中苯、甲苯、二甲苯,热解吸后进样,经 PEG-6000 柱分离后,用氢焰离子化监测器检测,以保留时间定性,峰高定量	苯、甲苯、二甲苯的检出限分别为 0.5×10^{-3}、1×10^{-3}、$2\times10^{-3}\mu g$(进样 $1\mu L$ 液体样品)	解吸方便,频率高
邻苯二甲酸二壬酯-有机皂土柱	苯、甲苯、二甲苯气样在-78℃浓缩富集,经邻苯二甲酸二壬酯及有机皂土色谱柱分离,用氢焰离子化检测器测定	苯、二甲苯的检出限分别为 $0.4mg/m^3$、$1.0mg/m^3$(1mL气样)	样品不稳定,需尽快分析

三、实验仪器与试剂

（一）仪器

（1）气相色谱仪，配有氢火焰离子化检测器。

（2）色谱柱：2m×3mm 的不锈钢柱，柱内填充涂覆 2.5%Bentane 的 Chromosorb W HPDMCS（80～100 目）。

（3）空气采样器：流量 0.2～1L/min。

（4）微量注射器：1 支，10μL。

（5）容量瓶：5mL、100mL 各 10 个。

（6）吸管：若干，1～20mL。

（7）采样管：15 支长 15cm、内径 8mm 的玻璃管，内装 20～50 目粒状活性炭 0.6g（活性炭预先在马弗炉内，经 350℃灼烧 3h，放冷后备用），或使用前在 300～350℃用氮气吹 10min，分成 A、B 两段，中间用玻璃棉隔开，两端密封保存。

（二）试剂

（1）苯（色谱纯试剂）。

（2）二硫化碳（CS_2）：使用前必须纯化（纯化方法见"注意事项"），并经色谱检验。进样 5μL，在苯与甲苯峰之间不出峰方可使用。

（3）苯标准贮备液：在 100mL 容量瓶中先加入 90mL 纯化后的 CS_2 溶液，然后加入 10μL 苯溶液，再用 CS_2 溶液定容，获得浓度为 88μg/mL 的苯标液。在 100mL 容量瓶中先加入 80mL CS_2 溶液，然后加入 10mL 浓度为 88μg/mL 的苯标液，再用 CS_2 溶液定容，获得浓度为 8.8μg/mL 的苯标准贮备液。此贮备液在 4℃可保存 1 个月。

四、实验步骤

（一）采样

用乳胶管连接采样管口与空气采样泵的进气口，并垂直放置，以 0.5L/min 流量，在窗口（室内侧）采样 40L 气体。采样后，用乳胶管将采样管两端套封，10 天内测定。记录采样点的气温和大气压力。

平行采样：另取两支管，按照以上的方法平行采样、测定，求出 3 次测量平均值。

室内其他位置采样：另取采样管，将其分别放置在门口（室内侧）、家具内、卧室内、客厅内设置采样点，用上述同样的方法采样并加以测定。

（二）测定

色谱条件：柱温为 64℃，气化室温度为 150℃，检测室温度为 150℃，载气（氮气）流量为 50mL/min，燃气（氢气）流量为 46mL/min，助燃气（空气）流量为 320mL/min。

标准曲线的绘制：分别取苯贮备液 0.0、5.0、10.0、15.0、20.0、25.0mL 于 100mL 容量瓶中，用 CS_2 稀释至标线，摇匀。另取 6 支 5mL 容量瓶，各加入 0.25g 粒状活性炭及 0～5 号的苯标液 2.00mL，振荡 2min，放置 20min 后，在上述色谱条件下，各进样 5.0μL，按所用气相色谱仪的操作要求测定标样的保留时间及峰高或峰面积。绘制峰高或峰面积与苯质量浓度之间关系的标准曲线。

样品的测定：将采样管两段活性炭分别移入 2 支 5mL 容量瓶中，加入纯化过的 CS_2

2.00mL，振荡15min，放置20min后，吸取5.0μL解吸液注入色谱仪，记录保留时间和峰高或峰面积。以保留时间定性，峰高或峰面积定量。

五、数据处理

空气中苯的质量浓度按公式（4-1-1）计算：

$$\rho = \frac{m_1 + m_2}{V_n} \tag{4-1-1}$$

式中 ρ——空气中苯的质量浓度，mg/m^3；
$\quad\quad m_1$——A段活性炭解吸液中苯系物的质量，μg；
$\quad\quad m_2$——B段活性炭解吸液中苯系物的质量，μg；
$\quad\quad V_n$——标准状态下的采样体积，L。

六、思考题

（1）测定空气中的苯有哪些方法？各自优缺点是什么？
（2）使用气相色谱法测定空气中的苯时应注意从哪些方面减少测定误差？
（3）请比较室内不同采样地点所取样品中苯质量浓度的大小，说明室内不同通风条件对苯质量浓度的影响。

七、注意事项

本法同样适用于空气中丙酮、苯乙烯、乙酸乙酯、乙酸丁酯、乙酸戊酯的测定。在以上色谱条件下，其保留时间见表4-1-4。

表 4-1-4 各组分的保留时间

组分	丙酮	乙酸乙酯	苯	甲苯	乙酸丁酯	苯乙烯
保留时间/s	0.65	0.76	1.00	1.89	2.53	6.94
组分	对二甲苯	间二甲苯	邻二甲苯	乙酸戊酯	乙苯	
保留时间/s	3.80	4.35	5.01	5.55	3.50	

（1）空气中苯质量浓度在$0.1mg/m^3$左右时，可用100mL注射器采气样，气样在常温下浓缩后，再加热解吸，用气相色谱法测定。
（2）市售活性炭、玻璃棉须经空白检验后，方能使用。检验方法是取用量为一支活性炭吸附采样管的玻璃棉和活性炭（分别约为0.1g和0.5g），经加纯化过的CS_2 2mL，振荡2min，放置20min，进样5μL，观察待测物位置是否有干扰峰。无干扰峰时方可使用，否则要预先处理。
（3）市售分析纯CS_2常含有少量苯与甲苯，须纯化后才能使用。纯化方法：1mL甲醛与100mL浓硫酸混合。取500mL分液漏斗一支，加入市售CS_2 250mL和甲醛-浓硫酸萃取液20mL，振荡分层。经多次萃取至CS_2呈无色后，再用20%Na_2CO_3水溶液洗涤两次，重蒸馏，截取46～47℃馏分。

实验三 室内空气中甲醛的浓度水平

甲醛是无色、具有强烈刺激性气味的气体，能与蛋白质结合。人吸入高浓度甲醛后，会严重刺激呼吸道和出现水肿、眼刺痛、头痛，也可能会发生支气管哮喘。经常吸入少量甲醛，能引起慢性中毒，出现黏膜充血、皮肤刺激以及头痛、心悸、失眠等症状，严重的还可导致癌症和遗传疾病。

甲醛是室内空气中常见的有机污染物。建筑和装修时使用的黏合剂、化纤地毯、油漆涂料以及化妆品、清洁剂、杀虫剂、食物和烹饪过程中等均可能挥发产生甲醛。室内空气中甲醛污染主要来自于装修，因室内装修及家具中甲醛等挥发物引起的恶性伤害案件经常会发生。人们约有80%的时间在室内度过，室内空气污染对人体健康造成的危害更大。因此，掌握室内空气中甲醛的分析方法、评价方法及其污染状况具有重要意义。

一、实验目的

（1）了解分光光度法测定空气中甲醛的基本原理，掌握其操作步骤。
（2）掌握室内空气中甲醛浓度水平的检测以及评价的基本方法。

二、实验原理

空气中的甲醛与酚试剂反应生成嗪，嗪在酸性溶液中被高价铁离子氧化形成蓝绿色化合物。用分光光度法测定其吸光度，可用于甲醛的定量分析。反应方程式如下：

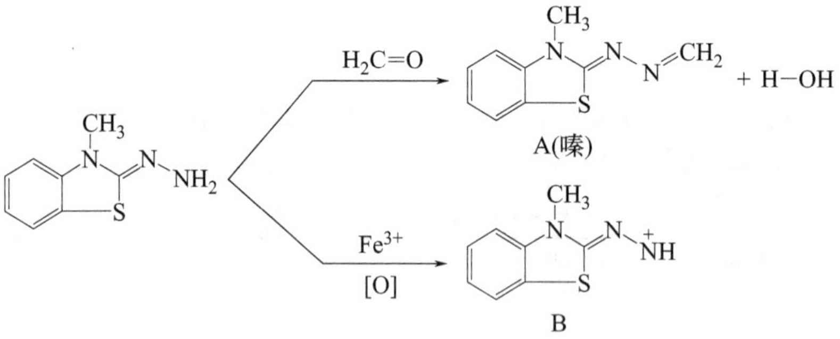

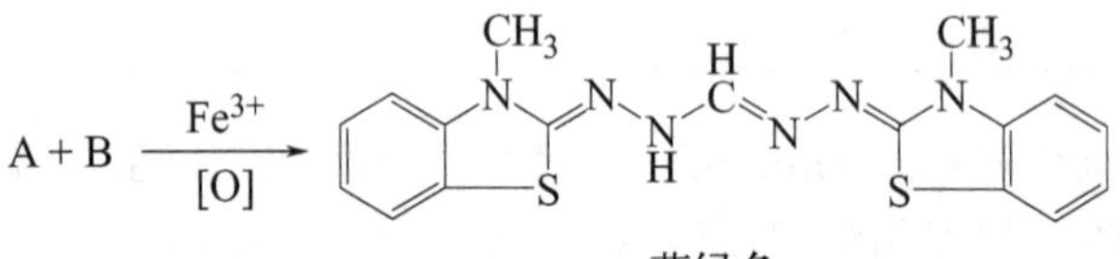

用 5mL 样品溶液，本方法的测定范围为 0.1～0.5μg；采样体积为 10L 时，可测质量浓度范围为 0.01～0.15mg/m³。灵敏度为每单位吸光度 2.8μg。甲醛检出下限为 0.056μg。当甲醛含量为每 5mL 0.4～1.0μg 时，样品加标回收率为 93%～101%。

二氧化硫共存时，测定结果偏低。因此，不可忽略二氧化硫的干扰，可将气体样品先通过硫酸锰滤纸过滤器，排除二氧化硫。

硫酸锰滤纸的制备：取 10mL 质量浓度为 100mg/mL 的硫酸锰水溶液，滴加到 250cm² 玻璃纤维滤纸上，风干后切成碎片，装入 1.5cm×150mm 的 U 形玻璃管中。采样时，将此管接在甲醛吸收管之前。此法制成的硫酸锰滤纸，有吸收二氧化硫的效能，受大气湿度影响很大，当相对湿度大于88%，采气速度为1L/min，二氧化硫浓度为1mL/m³ 时，能消除95%以上的二氧化硫，此滤纸可维持50h有效。当相对湿度为15%～35%时，其吸收二氧化硫的效能逐渐降低。所以相对湿度很低时，应换用新制的硫酸锰滤纸。

三、实验仪器与试剂

（一）仪器

(1) 大型气泡吸收管：出气口内径为 1mm，出气口至管底距离小于或等于 5mm。

(2) 恒流采样器：流量范围为 0~1L/min，流量稳定可调，恒流误差小于 2%，采样前和采样后应用皂沫流量计校准采样系统流量，误差小于 5%。

(3) 具塞比色管：10mL。

(4) 分光光度计：在 630nm 测定吸光度。

（二）试剂

本法中所用水均为重蒸馏水或去离子水，所有试剂一般为分析纯。

(1) 吸收液原液：称量 0.1g 酚试剂 [$C_6H_4SN(CH_3)C:NNH_2 \cdot HCl$，简称 MBTH]，加水溶解，转移到 100mL 具塞量筒中，加水至刻度，为吸收原液。放入 4℃ 冰箱中保存，可稳定 3d。实验时量取吸收原液 5mL，加 95mL 水，即为吸收液。使用前需重新配制。

(2) 甲醛标准储备液：取 2.8mL 质量分数为 36%~38% 甲醛溶液，放入 1L 容量瓶中，加水稀释至刻度。此溶液 1mL 约相当于含有 1mg 甲醛。其准确浓度需用碘量法标定。

甲醛标准储备液的标定：

① 10g/L 硫酸铁铵溶液：称量 1.0g 十二水合硫酸铁铵 [$NH_4Fe(SO_4)_2 \cdot 12H_2O$]，用 0.1mol/L 盐酸溶解，并稀释至 100mL。

② 碘溶液 [$c(1/2 I_2)=0.1000mol/L$]：称量 40g 碘化钾，溶于 25mL 水中，加入 12.7g 碘。待碘完全溶解后，用水定容至 1000mL。移入棕色瓶中，暗处储存。

③ 1mol/L 氢氧化钠溶液：称量 40g NaOH，溶于水中，并稀释至 1000mL。

④ 硫酸溶液 [$c(1/2 H_2SO_4)=0.5mol/L$]：取 28mL 浓硫酸缓慢加入水中，冷却后，稀释至 1000mL。

⑤ 硫代硫酸钠标准溶液，市售标准溶液。

⑥ 5g/L 淀粉溶液：将 0.5g 可溶性淀粉，用少量水调制成糊状，再加入 100mL 沸水，并煮沸 2~3min 至溶液透明。冷却后，加入 0.1g 水杨酸或 0.4g 氯化锌保存。

精确量取 20.00mL 待标定的甲醛标准储备液，置于 250mL 碘量瓶中。加入 20.00mL 碘溶液和 15mL 1mol/L 的氢氧化钠溶液，放置 15min，加入 20mL 硫酸溶液，再放置 15min，用硫代硫酸钠溶液滴定，至溶液呈现淡黄色时，加入 1mL 5g/L 的淀粉溶液继续滴定至恰使蓝色褪去为止，记录所用硫代硫酸钠溶液体积。同时用水作试剂空白滴定，记录空白滴定所用硫代硫酸钠标准溶液的体积。甲醛溶液的质量浓度用式（4-1-2）计算：

$$\rho_{甲醛}=\frac{(V_1-V_2)\times c \times 15}{20} \tag{4-1-2}$$

式中　$\rho_{甲醛}$——甲醛标准储备液质量浓度，mg/mL；

　　　V_1——试剂空白消耗硫代硫酸钠溶液的体积，mL；

　　　V_2——甲醛标准储备液消耗硫代硫酸钠溶液的体积，mL；

　　　c——硫代硫酸钠溶液的准确浓度，mol/L。

平行滴定两次，误差应小于 0.05mL，否则重新标定。

(3) 甲醛标准溶液：临用前，将甲醛标准储备液用水稀释，使 1.00mL 含 10μg 甲醛，立即再取此溶液 10.00mL，移入 100mL 容量瓶中，加 5mL 吸收原液，用水定容至 100mL，此液

1.00mL 含 1.00μg 甲醛，放置 30min 后，用于配制标准色列管。此标准溶液可稳定 24h。

四、实验步骤

（一）采样点的选择

根据所选择的室内环境的实际情况，采样前需要进行规划布点。采样点选择应遵循以下原则。

（1）采样点的数量：采样点的数量根据监测室内面积大小和现场情况而确定。原则上小于 50m² 的房间应设 1~3 个点；50~100m² 设 3~5 个点；100m² 以上至少设 5 个点。在对角线上或按梅花式均匀分布。

（2）采样点应避开通风口，离墙壁距离应大于 0.5m。

（3）采样点的高度：原则上与人的呼吸高度应一致，相对高度 0.5~1.5m。

（二）样品采集

用一个内装 5mL 吸收液的大型气泡吸收管，以 0.5L/min 流量，采气 10L。记录采样点的温度和大气压力。采样后样品在室温下应在 24h 内分析。

（三）绘制标准曲线

取 10mL 具塞比色管，用甲醛标准溶液按表 4-1-5 制备甲醛标准系列。

表 4-1-5 甲醛标准系列

管号	0	1	2	3	4	5	6	7	8
标准工作液体积/mL	0.00	0.10	0.20	0.40	0.60	0.80	1.00	1.50	2.00
吸收液体积/mL	5.0	4.9	4.8	4.6	4.4	4.2	4.0	3.5	3.0
甲醛质量/μg	0.0	0.1	0.2	0.4	0.6	0.8	1.0	1.5	2.0

各管中，加入 10g/L 硫酸铁铵溶液 0.4mL，摇匀。放置 20min。用 1cm 比色皿，在波长 630nm 下，以水作为参比，测定各管溶液的吸光度。以甲醛质量为横坐标，吸光度为纵坐标，绘制曲线，并计算回归斜率，以斜率倒数作为样品测定的计算因子 B_s（μg/吸光度）。

（四）样品测定

采样后，将样品溶液全部转入比色管中，用少量吸收液润洗吸收管，合并使总体积为 5mL。按绘制标准曲线的操作步骤测定吸光度（A）；在每批样品测定的同时，用 5mL 未采样的吸收液作试剂空白，测定试剂空白的吸光度（A_0）。

五、数据处理

（1）将采样体积按式（4-1-3）换算成标准状态下的采样体积：

$$V_0 = V_t \times \frac{T_0}{273+t} \times \frac{P}{P_0} \qquad (4-1-3)$$

式中 V_0——标准状态下的采样体积，L；

V_t——采样体积，即采样流量（L/min）× 采样时间（min）；

t——采样点的气温，℃；

T_0——标准状态下的热力学温度，K；

P——采样点的大气压力，kPa；

P_0——标准状态下的大气压力，101.3kPa。

(2) 空气中甲醛质量浓度按照式（4-1-4）计算：

$$\rho = \frac{(A-A_0)B_s}{V_0} \tag{4-1-4}$$

式中 ρ——空气中甲醛质量浓度，mg/m^3；
A——样品溶液的吸光度；
A_0——空白溶液的吸光度；
B_s——由实验计算得到的计算因子，$\mu g/$吸光度；
V_0——换算成标准状态下的采样体积，L。

六、思考题

(1) 试分析讨论室内空气中甲醛的主要来源。
(2) 了解甲醛的其他监测分析方法，与酚试剂分光光度法相比，有何优缺点？

实验四 环境空气中 SO_2 液相氧化模拟

一、实验目的

(1) 了解 SO_2 液相氧化的过程。
(2) 掌握 pH 法间接考查 SO_2 液相氧化过程的方法。

二、实验原理

SO_2 液相氧化的过程是大气降水酸化的主要途径。首先 SO_2 溶解于水中并发生一级和二级电离，生成 $SO_2 \cdot H_2O$、HSO_3^-、SO_3^{2-} 及 H^+，溶解态硫（Ⅳ）形态分布如图 4-1-1 所示。

溶解总硫的存在形式不仅与 SO_2 浓度有关，也与液相 pH 有关。一般条件下，典型大气液滴的 pH 为 2～6，此时 HSO_3^- 为溶解 S（Ⅳ）的主要存在形式，然后，溶解态的 S（Ⅳ）被氧化为 S（Ⅵ），常见的液相氧化剂包括 O_2、O_3、H_2O_2 和自由基等，其中溶解在水中的 O_2 是最常见也是最主要的氧化剂。在 SO_2 被 O_2 氧化的过程中，Fe（Ⅲ）和 Mn（Ⅱ）都可以起到催化剂的作用。

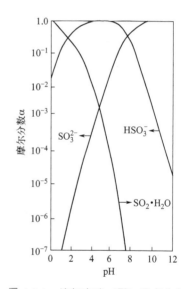

图 4-1-1 溶解态硫（Ⅳ）形态分布

$$Mn^{2+} + SO_2 = MnSO_2^{2+}$$
$$2MnSO_2^{2+} + O_2 = 2MnSO_3^{2+}$$
$$MnSO_3^{2+} + H_2O = Mn^{2+} + 2H^+ + SO_4^{2-}$$

总反应为

$$2SO_2 + 2H_2O + O_2 = 2SO_4^{2-} + 4H^+$$

水中的 Fe（Ⅲ）和 Mn（Ⅱ）主要来源于大气中的尘埃等杂质。

由于大气液滴中的 S（Ⅳ）主要以 HSO_3^- 的形式存在，因此在本实验中以 Na_2SO_3 溶液代替吸收了 SO_2 的液滴，模拟研究不同条件下 S（Ⅳ）的液相氧化过程。由于在 SO_3^{2-} 被

氧化为 SO_4^{2-} 的过程中溶液的 H^+ 浓度增加，pH 下降，因此本实验通过测定溶液的 pH 变化，估算 SO_2 的液相氧化速率。同时添加不同的催化剂，比较不同催化剂的催化效果。在本实验中，分别用 $MnSO_4$ 模拟 Mn（Ⅱ），用 $NH_4Fe(SO_4)_2$ 模拟 Fe（Ⅲ），用降尘和煤灰模拟实际大气液滴中的尘埃等各种杂质。

三、实验仪器与试剂

（一）仪器

(1) 精密 pH 计 2 台。

(2) 磁力搅拌器 6 台。

(3) 小型气泵。

(4) 2L 烧杯 1 个。

(5) 250mL 烧杯 6 个。

(6) 1L 容量瓶 3 个。

（二）试剂

(1) 亚硫酸钠溶液：0.01mol/L。溶解 1.26g 无水 Na_2SO_3 于水，定容到 1L。

(2) 硫酸锰溶液：0.0005mol/L。溶解 0.141g 无水 $MnSO_4$ 于烧杯中，用稀硫酸调节 pH＝5，转移到 1L 容量瓶中，定容。

(3) 硫酸铁铵溶液：0.0005mol/L。取 0.241g $NH_4Fe(SO_4)_2 \cdot 12H_2O$ 于烧杯中，加少量 1:4 的稀硫酸和适量水溶液，转移到 1L 容量瓶中，定容。使用时取适量溶液，用 NaOH 溶液小心调节 pH 至 5（注意避免沉淀）。

(4) 降尘-水悬浊液：收集并称取 0.2g 大气降尘（可取自室外窗台等处），放入 50mL 烧杯中，加 30mL 二次水，搅拌，并用稀硫酸调节 pH＝5。

(5) 煤灰-水悬浊液：称取 0.1g 煤灰，放入 50mL 烧杯中，加 30mL 二次水，搅拌，并用稀硫酸调节 pH＝5。

(6) 稀释水：取二次水 1.5L 于 2L 烧杯中，通空气 30min，同时用磁力搅拌器搅拌，最后用稀硫酸调节 pH 等于 5。

(7) 稀硫酸溶液：0.01mol/L。

(8) 稀氢氧化钠溶液：0.01mol/L。

(9) 标准缓冲溶液：0.05mol/L 邻苯二甲酸氢钾（pH＝4.01）及 0.025mol/L KH_2PO_4-0.025mol/L Na_2HPO_4（pH＝6.86）。

四、实验步骤

（一）模拟实验准备

(1) 取 250mL 烧杯 6 个，编号 1～6，分别用于模拟不加催化剂、加锰催化剂、加铁催化剂、加铁锰催化剂、加降尘催化剂和加煤灰催化剂 6 种情况。

(2) 向 1～4 号烧杯各加稀释水 190mL、0.01mol/L Na_2SO_3 溶液 10mL；向 5、6 号烧杯各加稀释水 160mL，0.01mol/L Na_2SO_3 溶液 10mL。

(3) 迅速向 2～6 号烧杯中依次加入以下试剂：1 号，去离子水 2mL；2 号，0.0005mol/L $MnSO_4$ 溶液 2mL；3 号，0.0005mol/L $NH_4Fe(SO_4)_2$ 溶液 2mL；4 号，

0.0005mol/L $MnSO_4$ 溶液和 0.0005mol/L $NH_4Fe(SO_4)_2$ 溶液各 1mL；5 号，降尘-水悬浊液 32mL；6 号，煤灰-水悬浊液 32mL。

(4) 加完各种试剂后，将 6 个烧杯置于磁力搅拌器上持续搅拌，用稀 H_2SO_4 和稀 NaOH 溶液迅速调节各烧杯 pH 至 5.0，并开始计时。

（二）液相氧化过程

每隔一定时间（5、10、15、20、25、30、40、50、60、70min）测定并记录各烧杯中溶液 pH 的变化。

五、数据处理

以 pH 为纵坐标、时间为横坐标绘制各体系中溶液 pH 随时间的变化曲线。评价并对比不同体系氧化反应的快慢，分析和对比各催化剂的催化作用。

六、思考题

(1) 为什么通过 pH 的变化可以估算 SO_2 液相氧化速率？
(2) 哪些因素会影响 SO_2 的氧化速率？
(3) 本实验成功的关键是什么？

实验五　空气中 SO_2 的测定

一、实验目的

(1) 了解二氧化硫比色测定的原理和方法。
(2) 熟悉并掌握空气污染物样品的采集技术。

二、实验原理

用盐酸副玫瑰苯胺法测定大气中二氧化硫是国内外广泛采用的较为成熟的方法。方法灵敏，选择性好，可用于短时间采样（例如 20～30min）或长时间采样（例如 24h），但吸收液毒性大。

二氧化硫被四氯汞钾溶液吸收后，生成稳定的二氯亚硫酸盐配合物，再与甲醛及盐酸副玫瑰苯胺作用，生成紫红色配合物，比色测定。主要干扰物质为氮氧化物、臭氧、锰、铁、铬等。加入氨基磺酸铵可消除氮氧化物的干扰，采样后放置一段时间可使臭氧自行分解，加入磷酸和乙二胺四乙酸二钠盐可以消除或减小某些重金属的干扰。环境大气中微量氨、硫化物及醛类不干扰。当采样体积为 30L 时，二氧化硫的最低检出质量浓度为 $0.025mg/m^3$。在质量浓度低于 $0.025mg/m^3$ 时，需增大采样体积，但必须检查及校正采样的吸收效率。二氧化硫质量浓度高于 $0.025mg/m^3$ 时，吸收效率大于 98%。

三、实验仪器与试剂

（一）仪器

(1) 吸收管：多孔玻板吸收管、小型冲击式吸收管或大型气泡吸收管，用于 30min 到 1h 采样；125mL 多孔玻板吸收瓶或 125mL 洗气瓶，用于 24h 采样。

(2) 大气采样器：流量范围 0～1L/min。
(3) 分光光度计。
(4) 恒温水浴器。

（二）试剂

(1) 0.04mol/L 四氯汞钾（TCM）吸收液：称取 10.9g 二氯化汞（$HgCl_2$）、6.0g 氯化钾（KCl）和 0.066g 乙二胺四乙酸二钠盐（EDTA-2Na），溶于水中，稀释至 1L。此溶液 pH 值约为 4，利用酸度计用 0.01mol/L 氢氧化钠溶液调节 pH 至 5.2 左右，此试剂在密闭容器中贮存可以稳定 6 个月。

(2) 质量浓度为 0.6% 的氨基磺酸铵溶液：称取 0.6g 氨基磺酸铵（$H_2NSO_3NH_4$）溶于水中，稀释至 100mL，临时现配。

(3) 质量浓度为 0.2% 的甲醛溶液：量取 1.4mL 36%～38%（质量分数）的甲醛，溶于水中，稀释至 250mL，于冰箱中保存，可稳定 1.5 个月。

(4) 0.05mol/L 碘储备液：称取 12.7g 碘放入烧杯中，加入 40g 碘化钾（KI），加 25mL 水，搅拌至全部溶解后，用水稀释至 1L，贮于棕色试剂瓶中。

(5) 0.005mol/L 碘溶液：量取 50mL 0.05mol/L 碘储备液，稀释至 500mL，贮于棕色试剂瓶中。

(6) 淀粉指示剂：称取 0.2g 可溶性淀粉（可加 0.4g 二氯化锌防腐），用少量水调成糊状物，倒入 100mL 沸水中，继续煮沸直到溶液澄清。冷却后贮于试剂瓶中。

(7) 0.015mol/L 碘酸钾标准溶液：称取 3.2100g 碘酸钾（KIO_3，优级纯，110℃ 烘 2h），溶于水中，移入 1000mL 容量瓶，用水稀释至标线。

(8) 0.1mol/L 硫代硫酸钠贮备液：称取 25g 硫代硫酸钠（$Na_2S_2O_3 \cdot 5H_2O$）溶于 1L 新煮沸但已冷却的水中，加 0.2g 无水碳酸钠，贮于棕色试剂瓶中，放置一周后标定其浓度，若溶液呈现浑浊时，应该过滤。

标定方法：吸取 10.00mL 0.1mol/L 碘酸钾标准溶液，置于 250mL 碘量瓶中，加入 70mL 新煮沸但已冷却的水，加 1g 碘化钾，振摇至完全溶解后，再加 3.5mL 冰醋酸（或 10mL 1mol/L 盐酸溶液），立即盖好瓶塞，混匀。在暗处放置 5min 后，用 0.1mol/L 硫代硫酸钠溶液滴定至淡黄色，加 5mL 新配制的 0.2% 淀粉指示剂后，溶液呈蓝色，再继续滴定至蓝色刚刚消失。计算硫代硫酸钠溶液的浓度。

(9) 0.01mol/L 硫代硫酸钠溶液：取 50.00mL 标定过的 0.1mol/L 硫代硫酸钠溶液置于 500mL 容量瓶中，用新煮沸并冷却后的水稀释至标线。

(10) 二氧化硫标准溶液：先配制亚硫酸钠水溶液。称取 0.200g 亚硫酸钠（Na_2SO_3）及 0.010g 乙二胺四乙酸二钠盐，溶于 200mL 新煮沸并冷却后的水中，轻轻摇匀（避免振荡，以防充氧），放置 2～3h 后标定。此溶液相当于每毫升含 320～400μg 二氧化硫。

标定方法如下：

取 4 个 250mL 碘量瓶（A_1、A_2、B_1、B_2），分别加入 50.00mL 0.005mol/L 碘溶液。在 A_1、A_2 内各加入 25mL 水，在 B_1、B_2 瓶内各加入 25.00mL 上述亚硫酸钠水溶液，盖好瓶塞。

立即吸取 2.00mL 上述亚硫酸钠水溶液，加入一个已加有 40～50mL 四氯汞钾溶液的 100mL 容量瓶中，使生成稳定的二氯亚硫酸盐配合物。

用四氯汞钾溶液将 100mL 容量瓶中溶液稀释至标线，摇匀。

将 A_1、A_2、B_1、B_2 四瓶于暗处放置 5min 后，用 0.01mol/L 硫代硫酸钠标准溶液滴定至浅黄色，加 5mL 新配制的 0.2% 淀粉指示剂，继续滴定至蓝色刚刚消失。平行滴定所用硫代硫酸钠标准溶液体积之差应不大于 0.10mL。

100mL 容量瓶中二氧化硫的质量浓度：

$$\text{二氧化硫溶液质量浓度}(SO_2, \mu g/mL) = \frac{(A-B)N \times 32000 \times 2.00}{25.00 \times 100} \tag{4-1-5}$$

式中 A——空白滴定所用硫代硫酸钠标准溶液体积的平均值，mL；

B——样品滴定所用硫代硫酸钠标准溶液体积的平均值，mL；

N——硫代硫酸钠标准溶液的浓度，mol/L。

根据以上计算的二氧化硫浓度，再用吸收液稀释成 $2.0\mu g/mL$ 的二氧化硫的标准溶液。此溶液置于冰箱 $0 \sim 4°C$ 冷藏，一周内浓度不变。

(11) 盐酸副玫瑰苯胺（即对品红）的提纯：取正丁醇和 1mol/L 盐酸溶液各 500mL，放在 1L 分液漏斗中摇匀，使其互溶达到平衡。称 0.100g 对品红于小烧杯中，加约 30mL 平衡过的 1mol/L 盐酸溶液，搅拌，放置至完全溶解后，用 1mol/L 盐酸将溶液分数次洗入 250mL 分液漏斗，溶液总体积不得超过 50mL。加 100mL 平衡过的正丁醇，振摇数分钟，静置至两相分层后，将下层含有对品红的盐酸溶液转入另一分液漏斗，加 100mL 平衡过的正丁醇，再抽提。按此操作，每次用 50mL 正丁醇再重复抽提 6 次，保留水相，尽量避免损失，弃去有机相。最后将水相滤入 1 支 50mL 容量瓶中，并用 1mol/L 盐酸溶液稀释至标线。此对品红贮备液（质量浓度为 0.2%）为浅棕黄色，应符合以下条件：

对品红贮备液在醋酸-醋酸钠缓冲溶液中，在波长 540nm 处有最大吸收峰。测定吸收曲线的溶液按下法配制：吸取 1.00mL 提纯后的对品红贮备液于 100mL 容量瓶中，用水稀释至标线，摇匀。取此稀释液 5.00mL 于 50mL 容量瓶中，加 5.00mL 1mol/L 醋酸-醋酸钠缓冲溶液，用水稀释至标线，1h 后，测定吸收曲线。

试剂空白值对温度敏感，用贮备液配制的对品红使用液，按本操作方法在 22°C 温度下绘制标准曲线时，使用 1cm 比色皿，在波长 548nm 处测得试剂空白液的吸光度应不超过 0.170。

在上述条件下绘制的标准曲线斜率应为 (0.030 ± 0.002) 吸光度/$\mu g\ SO_2$。

(12) 质量浓度为 0.016% 的对品红使用液：吸取 20.00mL 0.2% 对品红贮备液于 250mL 容量瓶中，加 25mL 3mol/L 磷酸溶液，用水稀释至标线。至少放置 24h 方可使用。此溶液可稳定 9 个月以上。

(13) 1mol/L 盐酸溶液：量取 86mL 浓盐酸（密度为 $1.19g/cm^3$），用水稀释至 1L。

(14) 3mol/L 磷酸溶液：量取 41mL 浓磷酸（H_3PO_4，85%），用水稀释至 200mL。

(15) 1mol/L 醋酸-醋酸钠缓冲溶液：称取 13.6g 醋酸钠（$NaCH_3COO \cdot 3H_2O$），溶于水，转入 100mL 容量瓶，加 5.7mL 冰醋酸，用水稀释至标线。此溶液 pH 值为 4.7。

试剂所用水为除去氧化剂的蒸馏水。

四、实验步骤

（一）采样

采样时间为 30min 或 60min 时，用 10mL 吸收液，流量为 0.5L/min。测定 24h 平均浓度时，用 75~100mL 吸收液，以 0.2~0.3L/min 流量连续采样 24h。

（二）标准曲线的绘制

取 7 个 25mL 容量瓶，按表 4-1-6 配制标准曲线。

表 4-1-6 标准曲线

瓶号	0	1	2	3	4	5	6
SO_2 标准溶液（$2\mu g/mL$）体积/mL	0.00	0.50	2.00	4.00	6.00	8.00	10.00
四氯汞钾溶液体积/mL	10.00	9.50	8.00	6.00	4.00	2.00	0.00
二氧化硫质量/μg	0.0	1.0	4.0	8.0	12.0	16.6	20.0

在以上各瓶中分别加入 1.00mL 0.6% 氨基磺酸铵溶液，摇匀，再加入 2.00mL 0.2% 甲醛溶液、5.00mL 0.016% 对品红使用液，用新煮沸并已冷却的水稀释至标线，摇匀。当室温为 15～20℃时，显色 30min；室温为 20～25℃时，显色 20min；室温为 25～30℃时，显色 15min。用 1cm 比色皿，于波长 548nm 处，以水为参比，测定吸光度。（因试剂空白对温度敏感，易受分光光度计中温度的影响，故以水为参比）。

为了提高准确度，可使用恒温水浴，绘制标准曲线时的温度与测定样品时的温度之差应不超过±2℃。

用最小二乘法计算标准曲线的回归方程式：

$$Y = bX + a \quad (4-1-6)$$

式中 Y——$A - A_0$ 为标准溶液吸光度（A）与试剂空白液吸光度（A_0）之差；
X——二氧化硫质量，μg；
b——回归方程式的斜率；
a——回归方程式的截距。

（三）样品测定

样品中若有浑浊物，应离心分离去除。

采样时间为 30min 或 1h 的样品，可将吸收液移入 25mL 容量瓶中，再用约 5mL 水冲洗吸收管。测定 24h 平均浓度时，用吸收液将样品体积调整至 75mL 或 100mL 标线，吸取 10.00mL 样品溶液于 25mL 容量瓶。

样品放置 20min，以使臭氧分解。

每一批样品应测定试剂空白液和控制样品，以检查试剂的可靠性和操作的准确性。配制方法：吸取 10.00mL 四氯汞钾溶液于 25mL 容量瓶，配成试剂空白液；吸取 2.00mL 二氧化硫标准溶液（$2.0\mu g/mL$）于 25mL 容量瓶中，加 8.00mL 四氯汞钾溶液，配成控制样品。

在试剂空白液、控制样品及全部样品中，分别加入 1.00mL 0.6% 氨基磺酸铵溶液，摇匀，放置 10min 以去除氮氧化物的干扰，以下步骤同"标准曲线的绘制"。

如果测定样品时的温度和绘制标准曲线时的温度相差不超过 2℃，则二者的试剂空白吸光度相差不应超过 0.03。如超过此值，应重新绘制标准曲线。

如样品的吸光度在 1.0～2.0 之间，可用试剂空白液稀释，经数分钟再测定吸光度，使测得的吸光度值在 0.03～1.0 之间，但稀释倍数不要大于 6 倍。

五、数据处理

$$二氧化硫质量浓度(SO_2, mg/m^3) = \frac{[(A - A_0) - a]D}{bV_r} \quad (4-1-7)$$

式中 A——样品溶液吸光度；

A_0——试剂空白液吸光度；

b——回归方程式的斜率；

a——回归方程式的截距；

V_r——换算为参比状态（25℃，1.01×10^5Pa）下的采样体积，L；

D——稀释因子（30min、1h样品$D=1$，24h平均浓度$D=7.5$或10）。

六、注意事项

（1）温度对显色有影响，温度越高，空白值越大；温度高时发色快，退色亦快，所以最好使用恒温水浴控制显色温度，并根据室温决定显色温度和时间。

（2）对品红的提纯可以降低试剂空白的吸光度，提高方法的灵敏度。增加酸度虽然也可以降低试剂空白的吸光度，但方法的灵敏度也随之降低。

（3）因六价铬能使紫红色配合物褪色，产生负干扰，故应避免用硫酸-铬酸洗液洗涤玻璃器皿。若已用硫酸-铬酸洗液洗过，则需用（1+1）盐酸溶液浸洗，再用水充分洗涤，以将六价铬洗净。

（4）用过的容量瓶及比色皿应及时用酸洗涤，否则红色难以洗净。容量瓶用（1+4）盐酸溶液洗，比色皿用（1+4）盐酸加1/3体积乙醇的混合液洗涤。

（5）四氯汞钾溶液为剧毒试剂，使用时应小心，如溅到皮肤上，立即用水冲洗。

含四氯汞钾废液的处理方法：在每升废液中加约10g碳酸钠至中性，再加10g锌粒，于黑布罩下搅拌24h后，将上清液倒入玻璃缸，滴加饱和硫化钠溶液，至不再产生沉淀为止。弃去溶液，将沉淀物转入以适当的容器里。此方法可以除去废液中的99%的汞。

（6）配制亚硫酸钠溶液时，应加入少量的EDTA二钠盐。SO_3^{2-}被水中的溶解氧氧化为SO_4^{2-}时，受试剂及水中微量Fe^{3+}的催化，加EDTA二钠盐配合Fe^{3+}，可使SO_3^{2-}浓度较为稳定。

实验六　盐酸萘乙二胺分光光度法测定空气中氮氧化物含量

常见的氮氧化物有一氧化氮（NO，无色）、二氧化氮（NO_2，红棕色）、一氧化二氮（N_2O）、五氧化二氮（N_2O_5）等，其中除五氧化二氮常态下呈固体外，其他氮氧化物常态下都呈气态。作为空气污染物的氮氧化物（NO_x）常指NO和NO_2。

天然排放的NO_x，主要来自土壤和海洋中有机物的分解，属于自然界的氮循环过程。人为活动排放的NO_x，大部分来自化石燃料的燃烧过程，如汽车、飞机、内燃机及工业窑炉燃料的燃烧过程；也来自生产、使用硝酸的过程，如氮肥厂、有机中间体厂、有色及黑色金属冶炼厂等。据20世纪80年代初估计，全世界每年由于人类活动向大气排放的NO_x约5300万吨。NO_x对环境的损害作用极大，它既是形成酸雨的主要物质之一，也是形成大气中光化学烟雾的重要物质和消耗O_3的一个重要因子。氮氧化物可刺激肺部，使人较难抵抗感冒之类的呼吸系统疾病，呼吸系统有问题的人士如哮喘病患者，会较易受二氧化氮影响。

中国《环境空气质量标准》规定氮氧化物的环境标准分为二级，一级标准与二级标准一致，日平均最大浓度限值均为$100\mu g/m^3$，适用于自然保护区、风景名胜区和其他需要特殊保护的区域,以及居住区、商业交通居民混合区、文化区、工业区和农村地区。

因此,掌握空气中氮氧化物含量测定具有十分重要的意义。

一、实验目的

(1) 掌握 Saltzman 法测定空气中的氮氧化物含量的基本原理和操作步骤。

(2) 了解环境空气中氮氧化物浓度水平对空气质量的影响。

二、实验原理

氮氧化物指空气中以一氧化氮和二氧化氮形式存在的氮的氧化物(以 NO_2 计)。Saltzman 实验系数指用渗透法制备的二氧化氮校准用混合气体,在采气过程中被吸收液吸收生成的偶氮染料相当于亚硝酸根的量与通过采样系统的二氧化氮总量的比值。氧化系数指空气中的一氧化氮通过酸性高锰酸钾溶液氧化管后,被氧化为二氧化氮且被吸收液吸收生成偶氮染料的量与通过采样系统的一氧化氮的总量之比。

空气中的二氧化氮被串联的第一支吸收瓶中的吸收液吸收并反应生成粉红色偶氮染料。空气中的一氧化氮不与吸收液反应,通过氧化管时被酸性高锰酸钾溶液氧化为二氧化氮,被串联的第二支吸收瓶中的吸收液吸收并反应生成粉红色偶氮染料。生成的偶氮染料在波长 540nm 处的吸光度与二氧化氮的含量成正比。分别测定第一支和第二支吸收瓶中样品的吸光度,计算两支吸收瓶内二氧化氮和一氧化氮的质量浓度,二者之和即为氮氧化物的质量浓度(以 NO_2 计)。

三、实验仪器与试剂

(一)仪器

(1) 分光光度计。

(2) 空气采样器:便携式空气采样器流量范围为 0.1~1.0L/min,采样流量为 0.4L/min 时,相对误差小于±5%;恒温、半自动连续空气采样器的采样流量为 0.2L/min 时,相对误差小于±5%,能将吸收液温度保持在 20℃±4℃。采样连接管线为硼硅玻璃管、不锈钢管、聚四氟乙烯管或硅胶管,内径约为 6mm,尽可能短些,任何情况下不得超过 2m,配有朝下的空气入口。

(3) 吸收瓶:可装 10mL、25mL 或 50mL 吸收液的多孔玻板吸收瓶,液柱高度不低于 80mm。吸收瓶的玻板阻力、气泡分散的均匀性及采样效率按《环境空气 氮氧化物的测定 Saltzman 法》(GB/T 15436—1995)附录 A 检查。图 4-1-2 示出较为适用的两种多孔玻板吸收瓶。使用棕色吸收瓶或采样过程中吸收瓶外罩黑色避光罩。新的多孔玻板吸收瓶或使用后的多孔玻板吸收瓶,应用 (1+1) HCl 浸泡 24h 以上,用清水洗净。

(4) 氧化瓶:可装 5mL、10mL 或 50mL 酸性高锰酸钾溶液的洗气瓶,液柱高度不能低于 80mm。使用后,用盐酸羟胺溶液浸泡洗涤。图 4-1-3 示出了较为适用的两种氧化瓶。

(二)试剂

除非另有说明,分析时均使用符合国家标准或专业标准的分析纯试剂和无亚硝酸根的蒸馏水、去离子水或相当纯度的水。必要时,实验用水可在全玻璃蒸馏器中每升水加入 0.5g 高锰酸钾($KMnO_4$)和 0.5g 氢氧化钡 [$Ba(OH)_2$] 重蒸。

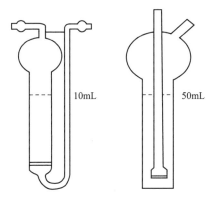
图 4-1-2　多孔玻板吸收瓶示意图

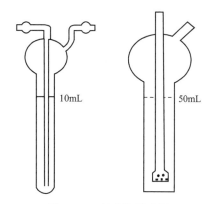
图 4-1-3　氧化瓶示意图

(1) 冰乙酸。

(2) 盐酸羟胺溶液，$\rho=0.2\sim0.5\text{g/L}$。

(3) 硫酸溶液，$c(1/2\ H_2SO_4)=1\text{mol/L}$。取 15mL 浓硫酸（$\rho=1.84\text{g/mL}$），徐徐加到 500mL 水中，搅拌均匀，冷却备用。

(4) 酸性高锰酸钾溶液，$\rho_{(KMnO_4)}=25\text{g/L}$。称取 25g 高锰酸钾于 1000mL 烧杯中，加入 500mL 水，稍微加热使其全部溶解，然后加入 1mol/L 硫酸溶液 500mL，搅拌均匀，贮于棕色试剂瓶中。

(5) N-(1-萘基)乙二胺盐酸盐 $[C_{10}H_7NH(CH_2)_2NH_2 \cdot 2HCl]$ 贮备液，$\rho=1.00\text{g/L}$。称取 0.50g N-(1-萘基)乙二胺盐酸盐于 500mL 容量瓶中，用水溶解稀释至刻度。此溶液贮于密闭的棕色瓶中，在冰箱中冷藏，可稳定保存三个月。

(6) 显色液。称取 5.0g 对氨基苯磺酸（$NH_2C_6H_4SO_3H$）溶解于约 200mL 40～50℃ 热水中，将溶液冷却至室温，全部移入 1000mL 容量瓶中，加入 50mL N-(1-萘基)乙二胺盐酸盐贮备溶液和 50mL 冰乙酸，用水稀释至刻度。此溶液贮于密闭的棕色瓶中，在 25℃ 以下暗处存放可稳定三个月。若溶液呈现淡红色，应弃之重配。

(7) 吸收液。使用时将显色液和水按 4∶1（体积分数）比例混合，即吸收液。吸收液的吸光度应小于等于 0.005。

(8) 亚硝酸盐（NO_2^-）标准贮备液，$\rho=250\mu\text{g/mL}$。准确称取 0.3750g 亚硝酸钠（$NaNO_2$，优级纯，使用前在 105℃±5℃ 干燥恒重）溶于水，移入 1000mL 容量瓶中，用水稀释至标线。此溶液贮于密闭棕色瓶中于暗处存放，可稳定保存三个月。

(9) 亚硝酸盐（NO_2^-）标准工作液，$\rho=2.5\mu\text{g/mL}$。准确移取亚硝酸盐标准贮备液 1.00mL 于 100mL 容量瓶中，用水稀释至标线。临用现配。

四、实验步骤

（一）样品采集

取两支内装 10.0mL 吸收液的多孔玻板吸收瓶和一支内装 5～10mL 酸性高锰酸钾溶液的氧化瓶（液柱高度不低于 80mm），用尽量短的硅胶管将氧化瓶串联在二支吸收瓶之间，以 0.4L/min 流量采气 4～24L。样品采集、运输及存放过程中避光保存，样品采集后尽快分析。

（二）标准曲线的绘制

取 6 支 10mL 具塞比色管，按表 4-1-7 制备亚硝酸盐标准溶液。根据表 4-1-7 分别移取相应体积的亚硝酸钠标准工作液，加水至 2.00mL，加入显色液 8.00mL。

表 4-1-7 NO_2^- 标准溶液

管号	标准工作液体积/mL	水体积/mL	显色液体积/mL	NO_2^- 质量浓度/(μg/mL)
0	0.00	2.00	8.00	0.00
1	0.40	1.60	8.00	0.10
2	0.80	1.20	8.00	0.20
3	1.20	0.80	8.00	0.30
4	1.60	0.40	8.00	0.40
5	2.00	0.00	8.00	0.50

各管混匀，于暗处放置 20min（室温低于 20℃时放置 40min 以上）。用 10mm 比色皿，在波长 540nm 处，以水为参比测量吸光度。扣除 0 号管的吸光度以后，对应 NO_2^- 的质量浓度（μg/mL），用最小二乘法计算标准曲线的回归方程。

标准曲线斜率控制在 0.960～0.978 吸光度·mL/μg。截距控制在 0.000～0.005 之间（以 5mL 体积绘制标准曲线时，标准曲线斜率控制在 0.180～0.195 吸光度·mL/μg，截距控制在±0.003 之间）。

空白试验：取实验室内未经采样的空白吸收液，用 10mm 比色皿，在波长 540nm 处，以水为参比测定吸光度。

（三）样品测定

采样后放置 20min，室温 20℃以下时放置 40min 以上。用水将采样瓶中吸收液的体积补充至标线，混匀。用 10mm 比色皿，在波长 540nm 处，以水为参比测量吸光度。同时，测定空白样品的吸光度。

若样品的吸光度超过标准曲线的上限，应用实验室空白试液稀释，再测定其吸光度。稀释倍数不得大于 6。

五、数据处理

(1) 空气中二氧化氮质量浓度 ρ_{NO_2}（mg/m³）按式（4-1-8）计算：

$$\rho_{NO_2} = \frac{(A_1 - A_0 - a)VD}{bfV_0} \tag{4-1-8}$$

(2) 空气中一氧化氮质量浓度 ρ_{NO}（mg/m³）以一氧化氮（NO）计，按式（4-1-9）计算：

$$\rho_{NO} = \frac{(A_2 - A_0 - a)VD}{bfV_0 K} \tag{4-1-9}$$

(3) 空气中氮氧化物的质量浓度 ρ_{NO_x}（mg/m³）以二氧化氮（NO₂）计，按式（4-1-10）计算：

$$\rho_{NO_x} = \rho_{NO_2} + \rho_{NO} \tag{4-1-10}$$

式中　A_1、A_2——串联的第一支和第二支吸收瓶中样品的吸光度；

　　　A_0——实验室空白的吸光度；

　　　b——标准曲线的斜率；

　　　a——标准曲线的截距；

　　　V——采样用吸收液体积，mL；

　　　V_0——换算为标准状态（101.325kPa，273K）下的采样体积，L；

　　　K——NO→NO_2 氧化系数，0.68；

　　　D——样品的稀释倍数；

　　　f——Saltzman 实验系数，0.88（当空气中二氧化氮质量浓度高于 0.72mg/m^3 时，f 取值 0.77）。

六、思考题

（1）试分析讨论环境空气中氮氧化物的主要来源。

（2）测定空气中氮氧化物时存在哪些干扰？如何消除？

实验七　靛蓝二磺酸钠分光光度法测定环境空气中的臭氧含量

臭氧是氧气的一种同素异形体，化学式是 O_3，分子量为 47.998，淡蓝色气体，液态为深蓝色，固态为紫黑色。气味类似鱼腥味，当浓度过高时，气味类似于氯气。臭氧有强氧化性，是比氧气更强的氧化剂，可在较低温度下发生氧化反应，在化工生产中可用臭氧氧化代替许多催化氧化或高温氧化，简化生产工艺并提高生产效率。液态臭氧还可用作火箭燃料的氧化剂。臭氧存在于大气中，靠近地球表面浓度为 0.001～0.03mg/L，这是由于大气中氧气吸收了太阳光波长小于 185nm 的紫外线。臭氧层可吸收太阳光中对人体有害的短波（300nm 以下）光线，防止这种短波光线射到地面，使生物免受紫外线的伤害。

臭氧的嗅阈值为 0.02mL/m^3，如果体积分数达到 0.1mL/m^3 时，就会刺激黏膜，达到 2mL/m^3 时会引起中枢神经障碍。美国国家环境空气质量标准（NAAQS）提出，人在一个小时内可接受臭氧的极限质量浓度是 260μg/m^3。在 320μg/m^3 臭氧环境中活动 1h 就会引起咳嗽、呼吸困难及肺功能下降。臭氧还能参与生物体中的不饱和脂肪酸、氨基及其他蛋白质反应，使长时间直接接触高浓度臭氧的人出现疲乏、咳嗽、胸闷胸痛等症状。

1951 年 A. J. 哈根最先指出臭氧（O_3）是氮氧化物、碳氢化合物和空气的混合物通过光化学反应形成的。之后 F. W. 温特发现 O_3 与不饱和烃（如汽车废气中的烃类）的化学反应产物跟洛杉矶烟雾有相同的伤害效应。因此，O_3 的浓度升高是光化学烟雾污染的标志。世界卫生组织已把臭氧的浓度水平作为判断大气环境质量的标准之一，并据此发布光化学烟雾的警报。

因此，掌握环境空气中臭氧含量测定具有十分重要的意义。

一、实验目的

（1）掌握靛蓝二磺酸钠分光光度法测定环境空气中的臭氧含量的基本原理和操作步骤。

（2）了解环境空气中臭氧浓度水平对空气质量的影响。

二、实验原理

空气中的臭氧在磷酸盐缓冲溶液存在下，与吸收液中蓝色的靛蓝二磺酸钠等物质的量反应，褪色生成靛红二磺酸钠，在610nm处测量吸光度，根据蓝色减退的程度定量空气中臭氧的质量浓度。

三、实验仪器与试剂

（一）仪器

（1）分光光度计：带20mm比色皿，可于波长610nm处测量吸光度。

（2）空气采样器：流量范围为0～1.0L/min，流量稳定。使用时，用皂膜流量计校准采样系统在采样前和采样后的流量，相对误差应小于±5%。

（3）多孔玻板吸收管：内装10mL吸收液，以0.50L/min流量采气，玻板阻力应为4～5kPa，气泡分散均匀。

（4）具塞比色管：10mL。

（5）生化培养箱或恒温水浴：温控精度为±1℃。

（6）水银温度计：精度为±0.5℃。

（二）试剂

除非另有说明，本实验所用试剂均为符合国家标准的分析纯化学试剂，实验用水为新制备的去离子水或蒸馏水。

（1）溴酸钾标准贮备溶液，$c(1/6\ KBrO_3)=0.1000mol/L$。准确称取1.3918g溴化钾（优级纯，180℃烘2h），置于烧杯中，加入少量水溶解，移入500mL容量瓶中，用水稀释至标线。

（2）溴酸钾-溴化钾标准溶液，$c(1/6\ KBrO_3)=0.0100mol/L$。移取10.00mL溴酸钾标准贮备溶液于100mL容量瓶中，加入1.0g溴化钾（KBr），用水稀释至标线。

（3）硫代硫酸钠标准贮备溶液，$c(Na_2S_2O_3)=0.1000mol/L$。

（4）硫代硫酸钠标准工作溶液，$c(Na_2S_2O_3)=0.0050mol/L$。临用前，取硫代硫酸钠标准贮备溶液用新煮沸并冷却到室温的水准确稀释20倍。

（5）硫酸溶液，1+6（体积比）。

（6）淀粉指示剂溶液，$\rho=2.0g/L$。称取0.20g可溶性淀粉，用少量水调成糊状，慢慢倒入100mL沸水，煮沸至溶液澄清。

（7）磷酸盐缓冲溶液，$c(KH_2PO_4\text{-}Na_2HPO_4)=0.050mol/L$。称取6.8g磷酸二氢钾（$KH_2PO_4$）、7.1g无水磷酸氢二钠（$Na_2HPO_4$），溶于水，稀释至1000mL。

（8）靛蓝二磺酸钠（$C_{16}H_{18}Na_2O_8S_2$）(简称IDS)，分析纯、化学纯或生化试剂。

（9）IDS标准贮备溶液。称取0.25g靛蓝二磺酸钠溶于水，移入500mL棕色容量瓶内，用水稀释至标线，摇匀，在室温暗处存放24h后标定。此溶液在20℃以下暗处存放可稳定2周。

标定方法：准确移取20.00mL IDS标准贮备溶液于250mL碘量瓶中，加入20.00mL溴酸钾-溴化钾溶液，再加入50mL水，盖好瓶塞，在16℃±1℃生化培养箱（或水浴锅）中放置至溶液温度与水浴温度平衡时，加入5.0mL硫酸溶液，立即盖塞、混匀并开始计时，于16℃±1℃水浴中暗处放置（35±1.0）min后，加入1.0g碘化钾，立即盖塞，轻轻摇匀至溶解，暗处放置5min，用硫代硫酸钠溶液滴定至棕色刚好褪去呈淡黄色，加入5mL淀粉指示剂溶液，继续滴定至蓝色消褪，终点为亮黄色。记录所消耗的硫代硫酸钠标准工作溶液的体积。

臭氧的质量浓度 ρ，由式（4-1-11）计算：

$$\rho=\frac{c_1V_1-c_2V_2}{V}\times 12.00\times 10^3 \tag{4-1-11}$$

式中 ρ——臭氧的质量浓度，$\mu g/mL$；

c_1——溴酸钾-溴化钾标准溶液的浓度，mol/L；

V_1——加入溴酸钾-溴化钾标准溶液的体积，mL；

c_2——滴定时所用硫代硫酸钠标准溶液的浓度，mol/L；

V_2——滴定时所用硫代硫酸钠标准溶液的体积，mL；

V——IDS 标准贮备溶液的体积，mL；

12.00——臭氧的摩尔质量（$1/4\ O_3$），g/mol。

（10）IDS 标准工作溶液：将标定后的 IDS 标准贮备液用磷酸盐缓冲溶液逐级稀释成每毫升相当于 $1.00\mu g$ 臭氧的 IDS 标准工作溶液。此溶液于 20℃ 以下暗处存放可稳定 1 周。

（11）IDS 吸收液：取适量 IDS 标准贮备液，根据空气中臭氧质量浓度的高低，用磷酸盐缓冲溶液稀释成每毫升相当于 $2.5\mu g$（或 $5.0\mu g$）臭氧的 IDS 吸收液。此溶液于 20℃ 以下暗处可保存 1 个月。

四、实验步骤

（一）样品采集与保存

用内装 $10.00mL\pm 0.02mL$ IDS 吸收液的多孔玻板吸收管，罩上黑色避光套，以 $0.5L/min$ 流量采气 $5\sim 30L$。当吸收液褪色约 60% 时（与现场空白样品比较），应立即停止采样。样品在运输及存放过程中应严格避光。当确信空气中臭氧的质量浓度较低，不会穿透时，可以用棕色玻板吸收管采样。样品于室温暗处存放至少可稳定 3d。用同一批配制的 IDS 吸收液，装入多孔玻板吸收管中，带到采样现场。除了不采集空气样品外，其他环境条件保持与采集空气的采样管相同。

（二）绘制标准曲线

取 10mL 具塞比色管 6 支，按表 4-1-8 制备标准溶液。

表 4-1-8 标准溶液

管号	IDS 标准溶液体积/mL	磷酸盐缓冲溶液体积/mL	臭氧质量浓度/($\mu g/mL$)
1	10.00	0.00	0.00
2	8.00	2.00	0.20
3	6.00	4.00	0.40
4	4.00	6.00	0.60
5	2.00	8.00	0.80
6	0.00	10.00	1.00

各管摇匀，用 20mm 比色皿，以水作参比，在波长 610nm 下测量吸光度。以校准系列中零浓度管的吸光度（A_0）与各标准溶液管的吸光度（A）之差为纵坐标，臭氧质量浓度为横坐标，用最小二乘法计算校准曲线的回归方程。

$$y = aX + b \tag{4-1-12}$$

式中 y——A_0-A,空白样品的吸光度与各标准溶液管的吸光度之差;

X——臭氧质量浓度,μg/mL;

b——回归方程的斜率;

a——回归方程的截距。

(三)样品测定

采样后,在吸收管的入气口端串接一个玻璃尖嘴,在吸收管的出气口端用吸耳球加压将吸收管中的样品溶液移入25mL(或50mL)容量瓶中,用水多次洗涤吸收管,使总体积为25.0mL(或50.0mL)。用20mm比色皿,以水作参比,在波长610nm下测量吸光度。

五、数据处理

空气中臭氧的质量浓度按式(4-1-13)计算:

$$\rho(O_3) = \frac{(A_0 - A - a)V}{bV_0} \tag{4-1-13}$$

式中 $\rho(O_3)$——空气中臭氧的质量浓度,mg/m³;

A_0——现场空白样品吸光度的平均值;

A——样品的吸光度;

b——标准曲线的斜率;

a——标准曲线的截距;

V——样品溶液的总体积,mL;

V_0——换算为标准状态(101.325kPa、273K)的采样体积,L。

所得结果精确至小数点后三位。

六、思考题

(1)试分析讨论环境空气中臭氧的主要来源。

(2)空气中的二氧化氮、二氧化硫对臭氧的测定结果会产生干扰吗?

第二节 水环境化学实验

实验八 饮用水中余氯的测定(碘量法滴定)

饮用水消毒过程中以液氯为消毒剂时,液氯与水中还原性物质或细菌等微生物作用之后,剩余在水中的残余氯量称为余氯,它包括游离性余氯和化合性余氯。

我国饮水厂的出厂水要求游离性余氯>0.3mg/L,管网水中游离性余氯>0.05mg/L。本实验采用碘量法测定水中余氯。

一、实验目的

(1)了解水中余氯测定的意义。

(2)掌握碘量法测定余氯的原理和操作。

二、实验原理

水中余氯在酸性溶液中与 KI 作用，释放出等化学计量的碘，以淀粉为指示剂，用 Na_2SO_3 标准溶液滴定至蓝色消失。由消耗的标准溶液的用量和浓度求出水中的余氯。主要反应为：

$$ClO^- + 2I^- + 2H^+ = I_2 + Cl^- + H_2O$$
$$I_2 + 2S_2O_3^{2-} = 2I^- + S_4O_6^{2-}$$

本法测定值为总余氯，包括 $HClO$、ClO^-、NH_2Cl 和 $NHCl_2$ 等。

三、实验仪器与试剂

（一）仪器

碘量瓶 300mL、滴定管 25mL、滴定台、容量瓶 500mL、烧杯 200mL、玻璃棒、胶头滴管、量筒 10mL。

（二）试剂

(1) 碘化钾：要求不含游离碘及碘酸钾。

(2) (1+5) 硫酸溶液。

(3) 重铬酸钾标准溶液，$c(1/6\ K_2Cr_2O_7) = 0.0250$ mol/L。称取 1.2259g 优级纯重铬酸钾（预先在 120℃下烘干 2h，在干燥器中冷却），溶于水中，移入 1000mL 容量瓶中，用水稀释至标线。

(4) 硫代硫酸钠标准溶液，$c(Na_2S_2O_3) = 0.05$ mol/L。称取 12.5g 硫代硫酸钠，$(Na_2S_2O_3 \cdot 5H_2O)$，溶于已煮沸放冷的水中，稀释至 1000mL。加入 0.2g 碳酸钠及数粒碘化汞，贮于棕色瓶内，溶液可保存数月。

硫代硫酸钠标准溶液的标定。移取 20.00mL 重铬酸钾标准溶液于碘量瓶中，加入 50mL 水和 1g 碘化钾，再加 5mL (1+5) 硫酸溶液，静置 5min 后，用待标定的硫代硫酸钠标准溶液滴定至淡黄色时，加入 1mL 1% 淀粉溶液，继续滴定至蓝色消失为止（注意：此时应带淡绿色，因为含有 Cr^{3+}），记录 $Na_2S_2O_3$ 标准溶液的用量。

硫代硫酸钠标准溶液浓度按式（4-2-1）计算：

$$c = \frac{c_1 \times 20.00}{V} \tag{4-2-1}$$

式中　c——硫代硫酸钠标准溶液的浓度，mol/L；

　　　c_1——重铬酸钾标准溶液浓度，mol/L；

　　　20.00——吸取重铬酸钾溶液的体积，mL；

　　　V——待标定硫代硫酸钠标准溶液用量，mL。

(5) 硫代硫酸钠标准滴定溶液，$c(Na_2S_2O_3) = 0.0100$ mol/L。移取上述已标定的 0.05mol/L 硫代硫酸钠标准溶液 100mL，移入 500mL 容量瓶中，用煮沸放冷的水稀释至刻度。

(6) 1% 淀粉溶液：称取 1.0g 可溶性淀粉以少量蒸馏水调成糊状，加入沸腾蒸馏水至 100mL，混匀。冷却后加入 0.1g 水杨酸或 0.4g 氯化锌作为防腐剂，防止指示剂腐败。

(7) 乙酸盐缓冲溶液（pH=4）：称取 146g 无水乙酸钠溶于水中，加入 457mL 乙酸，用水稀释至 1000mL。

四、实验步骤

（1）移取 100mL 水样（如水样中余氯质量浓度小于 1mg/L 时，可取 200mL 水样）于 300mL 碘量瓶内，加入 0.5g KI 和 5mL 乙酸盐缓冲溶液（调节 pH≈4）。

（2）用 0.0100mol/L $Na_2S_2O_3$ 标准滴定溶液滴定水样至变成淡黄色，加入 1mL 淀粉溶液，继续滴定至蓝色消失，记录 $Na_2S_2O_3$ 标准溶液用量。

五、数据处理

水中余氯按照式（4-2-2）计算：

$$总余氯(Cl_2)质量浓度(mg/L) = \frac{c_1 V_1 \times 35.453 \times 1000}{V_水} \quad (4-2-2)$$

式中　c_1——硫代硫酸钠标准滴定溶液浓度，mol/L；

　　　V_1——硫代硫酸钠标准滴定溶液用量，mL；

　　　$V_水$——水样体积，mL；

　　35.453——总余氯（$1/2Cl_2$）摩尔质量，g/mol。

数据记录与计算结果填入表 4-2-1 中：

表 4-2-1　$Na_2S_2O_3$ 标准溶液的标定及水样分析结果

	实验编号	1	2	3
$KMnO_4$ 标准溶液标定	取 $K_2Cr_2O_7$ 体积/mL			
	滴定管终读数/mL			
	滴定管始读数/mL			
	$Na_2S_2O_3$ 消耗体积/mL			
	$1/6K_2Cr_2O_7$ 浓度/(mol/L)			
	$Na_2S_2O_3$ 浓度/(mol/L)			
水样分析	取水样体积/mL			
	滴定管始读数/mL			
	滴定管终读数/mL			
	$Na_2S_2O_3$ 消耗体积/mL			
	总余氯(Cl_2)质量浓度/(mg/L)			

六、思考题

（1）饮用水出厂水和管网水中为什么必须含有一定量的余氯？

（2）滴定反应为什么必须在 pH≈4 的弱酸性溶液中进行？

实验九　水中 Cl^- 的测定（沉淀滴定）

氯离子（Cl^-）是水和废水中一种常见的无机阴离子。几乎所有的天然水中都有氯离子的存在，它的含量范围变化很大。在人类生存活动中，氯化物有很重要的生理作用及工业用途。

Cl^- 的测定采用沉淀滴定法，沉淀滴定法除必须符合滴定分析的基本要求外，还应满足：沉淀反应形成的沉淀溶解度必须很小；沉淀吸附现象不妨碍滴定终点的确定。

一、实验目的

（1）了解天然水和废水中氯离子的来源及含量。

（2）掌握沉淀滴定的原理和方法。

二、实验原理

在中性或弱碱性溶液中（pH=6.5~10.5），以铬酸钾（K_2CrO_4）为指示剂，用 $AgNO_3$ 标准溶液直接滴定水中 Cl^- 时，由于 AgCl 的溶解度（$8.72×10^{-8}$ mol/L）小于 Ag_2CrO_4 的溶解度（$3.94×10^{-7}$ mol/L），根据分步沉淀的原理，在滴定过程中，首先析出 AgCl 沉淀，沉淀反应为

$$Ag^+ + Cl^- \longrightarrow AgCl \downarrow$$
（白色沉淀）

当达到化学计量点后，水中 Cl^- 已全部滴定完毕，稍过量的 Ag^+ 便与 CrO_4^{2-} 生成 Ag_2CrO_4 砖红色沉淀，指示滴定终点到达。沉淀滴定反应为

$$2Ag^+ + CrO_4^{2-} \longrightarrow Ag_2CrO_4 \downarrow$$
（砖红色沉淀）

根据 $AgNO_3$ 标准溶液物质的量浓度和用量计算水样中 Cl^- 的含量。

三、实验仪器与试剂

（一）仪器

滴定管 50mL、滴定台、锥形瓶 25mL、胶头滴管、吸耳球、玻璃棒、烧杯 200mL。

（二）试剂

（1）NaCl 标准溶液的配制：将一定量 NaCl 放入坩埚中，于 500~600℃下熔化 40~50min。冷却后称取 8.2400g 用少量蒸馏水溶解，转移至 1000mL 容量瓶中，并稀释至刻度，该溶液浓度为 0.0141mol/L。移取 10mL，用水定容至 100mL，此溶液每毫升含 0.500mg 氯离子（Cl^-）。

（2）$AgNO_3$ 标准溶液（0.1000mol/L）的配制：称取 2.395g $AgNO_3$，溶于蒸馏水中并稀释至 1000mL。转入棕色试剂瓶中暗处保存，溶液浓度约为 0.0141mol/L。

（3）$AgNO_3$ 标准溶液的标定：用移液管分别移取 3 份 25mL 的 NaCl 标准溶液，同时移取 25mL 蒸馏水做空白，分别放入 250mL 锥形瓶中，各加 25mL 蒸馏水和 1mL K_2CrO_4 指示剂，在不断地摇动下用 $AgNO_3$ 溶液滴定至淡橘红色沉淀刚刚出现，即终点。记录 $AgNO_3$ 溶液的用量。根据 NaCl 标准溶液的浓度和 $AgNO_3$ 溶液的体积，计算 $AgNO_3$ 溶液的标准浓度，即

$$c_1 = \frac{c_2 V_2}{V_1} \tag{4-2-3}$$

式中　c_1——$AgNO_3$ 标准溶液的浓度，mol/L；

V_1——滴定 $AgNO_3$ 标准溶液的体积，mL；

c_2——NaCl 标准溶液的浓度，mol/L；

V_2——吸取 NaCl 标准溶液的体积，mL。

（4）K_2CrO_4 指示剂溶液（5%）的配制。称取 5g K_2CrO_4 溶于少量水中，用上述 $AgNO_3$ 溶液滴至有红色沉淀生成，混匀。静置 12h，过滤，滤液滤入 100mL 容量瓶中，用蒸馏水稀释至刻度。

四、实验步骤

（1）如果水样的 pH 值在 6.5～10.5 范围时，可直接滴定；超出此范围的水样应以酚酞作指示剂，用 0.05mol/L 的 H_2SO_4 溶液或 NaOH 溶液调节 pH≈8.0。

（2）水样中有机物含量高或色度大时，取 150mL 水样，放入 250mL 锥形瓶中，加 2mL 氢氧化铝悬浮液，振荡过滤，弃去最初滤液 20mL。

（3）如果水样中含有硫化物、亚硫酸盐或硫代硫酸盐，用氢氧化钠溶液调水样至中性或弱碱性，加 1mL 质量分数为 30% 的 H_2O_2，混匀。1min 后加热至 70～80℃，除去过量的 H_2O_2。

（4）如果水样中高锰酸盐指数（以 O_2 计）大于 15mg/L，则加入少量 $KMnO_4$，沸腾，再加数滴乙醇除去过量 $KMnO_4$，然后过滤取样。

（5）水样分析。吸取 50mL 水样 3 份和 50mL 蒸馏水（作空白实验）分别放入锥形瓶中；加入 K_2CrO_4 1mL 指示剂，在剧烈摇动下用 $AgNO_3$ 标准溶液滴定至刚刚出现淡橘红色，即终点。记录 $AgNO_3$ 标准溶液用量。

五、数据处理

根据下列公式计算：

$$\text{氯离子质量浓度}(\text{mg/L}) = \frac{(V_2 - V_0)c \times 35.453 \times 1000}{V_\text{水}} \tag{4-2-4}$$

式中　V_2——水样消耗 $AgNO_3$ 标准溶液的体积，mL；

c——$AgNO_3$ 标准溶液的浓度，mol/L；

V_0——蒸馏水消耗 $AgNO_3$ 标准溶液的体积，mL；

$V_\text{水}$——水样体积，mL；

35.453——Cl^- 的摩尔质量，g/mol。

实验结果及计算结果填入表 4-2-2 中。

表 4-2-2　$AgNO_3$ 标准溶液的标定及水样分析结果

	实验编号	1	2	3
溶液标定	取 NaCl 标准溶液的体积/mL			
	滴定终点读数/mL			
	滴定始点读数/mL			
	滴定 $AgNO_3$ 标准溶液的体积/mL			
	NaCl 标准溶液的浓度/(mol·L^{-1})			
	$AgNO_3$ 标准溶液的浓度/(mol·L^{-1})			

续表

水样测定	实验编号	1	2	3
	取水样体积/mL			
	滴定终点读数/mL			
	滴定始点读数/mL			
	滴定 $AgNO_3$ 标准溶液的体积/mL			
	氯离子(Cl^-)质量浓度/(mg·L^{-1})			

六、思考题

(1) 以莫尔法测定水中 Cl^- 时，为什么在中性或弱碱性溶液中进行？

(2) 以 K_2CrO_4 作指示剂时，指示剂浓度过高或过低对测定有何影响？

(3) 用 $AgNO_3$ 标准溶液滴定 Cl^- 时，为什么必须剧烈摇动？

实验十　Fenton 试剂催化氧化染料废水

一、实验目的

(1) 了解 Fenton 试剂的性质。

(2) 了解 Fenton 试剂降解有机污染物的机理。

(3) 掌握 Fenton 反应中各因素对废水脱色率的影响规律。

二、实验原理

Fenton 试剂的氧化机理可以用下面的化学反应方程式表示：

$$Fe^{2+} + H_2O_2 \longrightarrow Fe^{3+} + OH^- + OH·$$

OH·的生成使 Fenton 试剂具有很强的氧化能力。研究表明，在 pH=4 的溶液中，其氧化能力在溶液中仅次于氟气。因此，持久性有机污染物，特别是芳香族化合物及一些杂环类化合物，均可以被 Fenton 试剂氧化分解。本实验采用 Fenton 试剂法处理甲基橙模拟染料废水。配制一定浓度的甲基橙模拟废水，实验时取该废水于烧杯（或锥形瓶）中，加入一定量的硫酸亚铁，开启恒温磁力搅拌器，使其充分混合溶解，待溶解后，迅速加入设定量的 H_2O_2，混匀，反应至所设定时间，用 NaOH 溶液终止反应，调节 pH 值为 8~9，静置适当时间，取上层清液在最大吸收波长 465nm 处测吸光度，色度去除率=（反应前后最大吸收波长处的吸光度差/反应前的吸光度）×100%。

三、实验仪器与试剂

（一）仪器

(1) 酸度计或 pH 试纸。

(2) 722 可见光分光光度计。

(3) 托盘天平、分析天平、250mL 锥形瓶 15 个、2mL 吸量管 5 个、5mL 吸量管 4 个、100mL 的量筒 4 个。

（二）试剂

（1）甲基橙。

（2）$FeSO_4 \cdot 7H_2O$、H_2O_2（30%）、H_2SO_4、$NaOH$（均为分析纯）。

四、实验步骤

（1）配制 200mg/L 的甲基橙模拟废水：实验时，取 200mg/L 的甲基橙模拟废水 200mL 于烧杯（或锥形瓶）中。

（2）确定适宜的硫酸亚铁投加量。具体做法如下：甲基橙模拟废水的质量浓度为 200mg/L，H_2O_2（30%）投加量为 1mL/L，水样的 pH 值为 4.0~5.0，水样温度为室温时，投加不同量的 $FeSO_4 \cdot 7H_2O$（投加量分别为 20mg/L、60mg/L、100mg/L、200mg/L、300mg/L）进行脱色实验，反应时间为 90min。通过此实验，确定出 $FeSO_4 \cdot 7H_2O$ 的最佳投加量。

（3）确定适宜的 H_2O_2（30%）投加量。具体做法如下：甲基橙模拟废水的浓度为 200mg/L，$FeSO_4 \cdot 7H_2O$ 的投加量为（2）中确定的最佳投加量，水样的 pH 值为 4.0~5.0，水样温度为室温时，投加不同量的 H_2O_2（30%）（投加量分别为 0.1mL/L、0.2mL/L、0.4mL/L、0.6mL/L、0.8mL/L）进行脱色实验，反应时间为 90min。通过此实验，确定出 H_2O_2（30%）的最佳投加量。

（4）确定 pH 对降解效果的影响：甲基橙模拟废水的浓度为 200mg/L，$FeSO_4 \cdot 7H_2O$ 的投加量为（2）中确定的最佳投加量，H_2O_2（30%）投加量为（3）中确定的最佳投加量，考察 pH（pH 分别为 1、2、3、4、5、6）对甲基橙模拟废水降解效果的影响，确定最佳 pH。

（5）确定反应时间对降解效果的影响：甲基橙模拟废水的浓度为 200mg/L，水样的 pH 值为 4.0，$FeSO_4 \cdot 7H_2O$ 的投加量为（2）中确定的最佳投加量，H_2O_2（30%）投加量为（3）中确定的最佳投加量，在最佳 pH 条件下考察反应时间（取样时间分别为 10min、20min、40min、60min、120min）对甲基橙模拟废水降解效果的影响。

五、数据处理

$$色度去除率 = \frac{反应前后最大吸收波长处的吸光度差}{反应前的吸光度} \tag{4-2-5}$$

六、思考题

总结 Fenton 反应中各因素对废水脱色率的影响规律。

七、拓展内容

偶氮染料是构成工业用染料的最大一部分，甲基橙作为一种代表性的酸性偶氮染料，目前使用较为广泛，化学上用作试剂和指示剂，工业上用于对棉、麻、纸张及皮革等的染色。甲基橙具有结构稳定、难挥发、难生物降解等特性。甲基橙分子式为 $C_{14}H_{14}N_3NaO_3S$，分子量为 327.33，其化学结构见图 4-2-1。甲基橙染料呈弱酸性，最大吸收波长为 464nm，如图 4-2-2 所示。

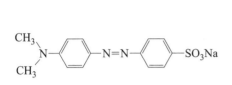

图 4-2-1　甲基橙化学结构式　　　　图 4-2-2　甲基橙紫外可见吸收波长谱图

实验十一　天然水中铜的存在形态

一、实验目的

（1）判别铜在湖水中几种简单的结合状态。
（2）学习用阳极溶出伏安法测定水中金属结合状态的一般实验技术。
（3）熟悉掌握 XJP-821（B）型新极谱仪的使用方法。

二、实验原理

天然水中重金属的存在形态，按其物理状态可分为颗粒态和溶解态两类。前者包括吸附、络合于悬浮物粒子上的各种化学态。后者按其在水中的活性又分为稳定态和不稳定态。不稳定态主要包括游离的金属离子、弱结合的有机和无机的络合吸附态金属。处于这一状态的金属有电活性，在电极上能反应。稳定态主要包括强结合的有机和无机络合吸附态金属。它们在电极上无反应，但经紫外光照射后，其中的有机结合态，即稳定态 A，会变成不稳定态。不被紫外光分解的部分，即稳定态 B，经硝酸-高氯酸消化后也会变成不稳定态。不稳定态有电活性，能被微电极富集，可用溶出伏安法测定。

在适当的底液及外加电压下，不稳定态铜可以还原为金属铜沉积在工作电极上。即不稳定态铜有电活性，可以用电解法富集。而稳定态与电极无反应，不能被富集。这是用电化学法判别天然水中重金属的稳定态与不稳定态的重要依据。再辅以紫外光照射、强酸消解等方法，还可把稳定态进一步分为 A、B 两态。富集在工作电极上的铜膜，当由负电位等速向正电位扫描过程中电位达到铜的溶出电位时，铜迅速氧化成铜离子，渗入溶液中，同时形成一个溶出电流峰。在其他条件不变时，可以根据溶出电流峰高度来确定被测液中铜的含量。

本实验测定受铜轻度污染的湖水中铜的结合状态。湖水用 $0.45\mu m$ 膜过滤，用 XJP-821（B）型新极谱仪和 ATA-1A 型旋转圆盘电极组成的极谱技术分别测定：

① 未加处理的滤液中的铜含量；
② 经紫外光照处理后的滤液中的铜含量；
③ 用硝酸和高氯酸消化后的滤液中的铜含量；
④ 经硝酸和高氯酸消化后的悬浮物中的铜含量。

显然，未加处理的滤液中被测出的只是不稳定态铜。经紫外光照射后的滤液中测出的是不稳定态铜加上稳定态 A 铜。经过硝酸和高氯酸消化后的滤液中测出的则包括不稳定态铜、稳定态 A 铜和稳定态 B 铜。因此，经过计算后可以分别确定不稳定态、稳定态 A、稳定态 B 和颗粒态的铜含量。

三、实验仪器与试剂

XJP-821（B）型新极谱仪、ATA-1A 型旋转圆盘电极、紫外灯照射器、电热板（800W）、微孔膜过滤器、真空泵系统、氮气钢瓶、移液管（0.25mL、2mL、5mL、50mL、100mL）、容量瓶 100mL、浓 $HClO_4$（GR）、浓 HNO_3（GR）、30% H_2O_2（GR）、铜标准使用液（1.00mg/L）、底液（3mol/L NH_4Cl-NH_4Ac-$NH_3 \cdot H_2O$）。

四、实验步骤

（1）水样采集和处理：用塑料桶取 1L 湖水，用玻璃纤维滤去大块颗粒物，再用 100mL 移液管取出 200mL 水样放入微孔膜过滤器中抽滤。弃去前面 10mL 滤液，余者存于冰箱中备用。小心移下微孔膜及悬浮物，放入 150mL 烧杯内备用。

（2）滤液光照处理：移取 50mL 滤液到紫外灯照射器内，再加 10 滴 30% H_2O_2，小心放入搅拌磁子。接通冷却水和电源，在搅拌中光照滤液 2h。照完后把滤液倒入 100mL 烧杯中备用。

（3）滤液消化处理：移取 50mL 滤液到 150mL 烧杯中，加 2mL 浓 HNO_3，在电热板上煮沸。试液近干时加 2mL 浓 $HClO_4$，继续加热到白烟将尽、内容物近干时为止。取下烧杯，冷却后加 10mL 蒸馏水，电热板上煮沸 1min。冷后移入 100mL 容量瓶内，用蒸馏水定容备用。

（4）滤膜处理：往盛有微孔膜和悬浮物的烧杯内加入 10mL 蒸馏水和 2mL 浓硝酸，然后继续按滤液消化处理过程操作。

（5）测定：移取溶液 50.0mL 于电解池杯中，再加 5.0mL 3mol/L NH_4Cl-NH_4Ac-$NH_3 \cdot H_2O$ 底液，将 ATA-1A 型旋转圆盘电极插入电解池内，调节电极转速，通高纯氮除氧 5 分钟。同时密封电解池系统，按下电极开关，触发富集键，2min 后仪器自动静止 30 秒扫描，得到一个溶出峰，测量峰高 h，重复测定 3~4 次（注意每次测定前须进行电极电化学清洗）。然后，在电解池杯中加入 0.25mL 1.00mg/L 的铜标准液，在相同条件下重复测定 3~4 次，记录峰高 H。阳极溶出伏安法测定条件见表 4-2-3。滤液光照试样、滤液消化试样、滤膜试样测定同上。

表 4-2-3 阳极溶出伏安法测定条件

上限电位/V	−1.0	富集时间/min	2
起始电位/V	−0.9	电极转速/(r/min)	1000
下限电位/V	0	X 轴量程/(mV/cm)	100
扫描速度/(mV/s)	100	Y 轴量程/(mV/cm)	50

五、数据处理

（1）根据公式（4-2-6）和（4-2-7）计算不同方式处理试样的铜含量，结果填入表 4-2-4。

$$c'_x = \frac{nc_0 h}{H(m+n) - hm} \qquad (4\text{-}2\text{-}6)$$

$$c_x = \frac{mc'_x}{V} \times 1000 \qquad (4\text{-}2\text{-}7)$$

式中　c'_x——被测试液+底液中铜的质量浓度，$\mu g/mL$；

　　　h——被测试液峰高，mm；

　　　H——加标后被测试液峰高，mm；

　　　V——被测试液体积，mL；

　　　m——被测试液体积 V+底液体积，mL；

　　　n——加入铜标准液体积，mL；

　　　c_0——加入铜标准溶液的质量浓度，mg/L；

　　　c_x——被测试液铜质量浓度，$\mu g/L$。

表 4-2-4　不同处理方式 Cu 含量测定

处理方式	直接处理	光照处理	消化处理	滤膜处理
被测液峰高 h/mm				
加标后被测液峰高 H/mm				
铜质量浓度/($\mu g/L$)				

（2）根据各种处理方式所得铜质量浓度算出湖水中颗粒态、不稳定态、稳定态 A、稳定态 B 的铜质量浓度，把结果填入表 4-2-5。

表 4-2-5　湖水中不同形态下的 Cu 质量浓度

形态	颗粒态	溶解态		
		不稳定态	稳定态 A	稳定态 B
铜质量浓度/($\mu g/L$)				

六、思考题

（1）为什么可用极谱技术来测定水中金属形态？

（2）简述阳极溶出伏安法灵敏度高的原因？

（3）为了保证测量准确，应注意哪些关键问题？

七、注意事项

（1）由于湖水来源不同，铜含量不同，因而移取试样的体积和加入标准溶液的体积可视具体情况作适当调整。

（2）在试液用 HNO_3 和 $HClO_4$ 消化时，消化完全后应加热至白烟将尽，但勿蒸干。如溶液中剩有过多的酸，则试液酸度很高，影响测定；如蒸干，则局部温度过高，微量的铜可能挥发损失。

（3）每经过一次溶出测定，需对电极进行电化学清洗，经扫描查验确无铜存在后再做下一次测定。

(4) 每个样品要取得重现性较好的溶出电流峰，必须注意使每次富集时间、静置时间、电极清洗时间等保持严格一致。

实验十二　废水中生化需氧量（BOD_5）的测定

一、实验目的

(1) 掌握水样的采集和处理方法。
(2) 掌握 BOD_5 的测定原理和操作。

二、实验原理

取两份生活污水，一份测定当时的溶解氧，另一份在（20±1）℃下培养5天再测定溶解氧，两者之差即为 BOD_5。

溶解氧的测定原理：在水样中加入硫酸锰和碱性碘化钾，二价锰先生成白色的 $Mn(OH)_2$ 沉淀，然后很快被水中溶解氧氧化为三价或四价的锰，从而将溶解氧固定。在酸性条件下，高价的锰可以将 I^- 氧化为 I_2，然后用硫代硫酸钠标准溶液滴定生成的 I_2，即可求出水中溶解氧的含量。

三、实验仪器与试剂

（一）仪器

(1) 恒温培养箱。
(2) 溶解氧瓶（200～300mL）：带有磨口玻塞，并具有供水封闭的钟形口。

（二）试剂

(1) 硫酸锰溶液：称取480g硫酸锰（$MnSO_4 \cdot H_2O$）溶于水，用水稀释至1000mL。此溶液加至酸化过的碘化钾溶液中，遇淀粉不得产生蓝色。
(2) 碱性碘化钾溶液：称取500g氢氧化钠溶解于300～400mL水中，另称取150g碘化钾溶于200mL水中，待氢氧化钠溶液冷却后，将两溶液合并，混匀，用水稀释至1000mL。如有沉淀，放置过夜后，倾出上层清液，储于棕色瓶中，用橡皮塞塞紧，避光保存。此溶液酸化后，遇淀粉应不呈蓝色。
(3) 硫代硫酸钠溶液：称取2.5g硫代硫酸钠（$Na_2S_2O_3 \cdot 5H_2O$）溶于煮沸放冷的水中，加0.2g碳酸钠，用水稀释至1000mL，储于棕色瓶中。使用前用重铬酸钾标准溶液标定。
(4) 浓硫酸（$\rho=1.84g/mL$）。
(5) 0.5%淀粉溶液：称取0.5g可溶性淀粉，用少量水调成糊状，再用刚煮沸的水稀释至100mL。冷却后，加入0.1g水杨酸和0.4g氯化锌防腐。

四、实验步骤

(1) 样品采集：准备好6个溶解氧瓶，用虹吸法把水样转移到溶解氧瓶内，并使水样从瓶口溢出数秒钟。其中3瓶固氧，并测定其溶解氧，另3瓶放在恒温培养箱中培养5天后再测定溶解氧。

(2) 溶解氧固定：用移液管插入溶解氧瓶的液面下，加入 1mL 硫酸锰溶液、2mL 碱性碘化钾溶液，盖好瓶塞，颠倒混合数次，静置。

(3) 溶解氧的测定：打开瓶塞，立即用移管插入液面下加入 2.0mL 浓硫酸。盖好瓶塞，颠倒混合摇匀，至沉淀物全部溶解，放于暗处静置 5min。

吸取 100.00mL 上述溶液于 250mL 锥形瓶中，用硫代硫酸钠标准溶液滴定至溶液呈淡黄色，加 1mL 淀粉溶液，继续滴定至蓝色刚好褪去，并记录硫代硫酸钠溶液用量。

五、数据处理

$$溶解氧(O_2)质量浓度(mg/L) = \frac{1}{4} \times \frac{cVM_{O_2}}{V_0} \times 10^3 \quad (4\text{-}2\text{-}8)$$

式中　c——硫代硫酸钠标准溶液浓度，mol/L；
　　　V——滴定消耗硫代硫酸钠标准溶液体积，mL；
　　　V_0——滴定时所取水样的体积，mL；
　　　M_{O_2}——O_2 的摩尔质量，g/mol。

六、注意事项

(1) 水样的pH值若超过 6.5~7.5 范围时，可用盐酸或氢氧化钠稀溶液调节至 7，但用量不要超过水样体积的 0.5%。

(2) 若从水温较低的水域中采集水样，可能含有过饱和溶解氧，此时应将水迅速升温至 20℃左右，充分振摇，以赶出过饱和的溶解氧。

若从水温较高的水浴或污水排放口取得的水样，则应迅速使其冷却至 20℃左右，并充分振摇，使其与空气中氧分压接近平衡。

实验十三　工业废水中铬的测定（二苯碳酰二肼分光光度法）

一、实验目的

(1) 掌握二苯碳酰二肼分光光度法测定水中六价铬和总铬的原理和方法。
(2) 学习用 Microsoft Office Excel 求线性回归方程的方法。

二、实验原理

在酸性溶液中，六价铬离子与二苯碳酰二肼反应，生成紫红色化合物，其最大吸收波长为 540nm，吸光度与浓度的关系符合比尔定律。如果测定总铬，需先用高锰酸钾将水样中的三价铬氧化为六价铬，再进行测定。

三、实验仪器与试剂

（一）仪器

分光光度计。

（二）试剂

(1) 丙酮。

(2) （1+1）硫酸。

(3) （1+1）磷酸。

(4) 2g/L 氢氧化钠溶液。

(5) 氢氧化锌共沉淀剂。称取硫酸锌（$ZnSO_4 \cdot 7H_2O$）8g，溶于 100mL 水中；称取氢氧化钠 2.4g，溶于 120mL 水中。将两溶液混合。

(6) 40g/L 高锰酸钾溶液。

(7) 铬标准贮备液。称取于 120℃ 干燥 2h 的重铬酸钾（GR）0.2829g，用水溶解，移入 1000mL 容量瓶中，用水稀释至标线，摇匀。

(8) 铬标准使用液。移取 5.00mL 铬标准贮备液于 500mL 容量瓶中，用水稀释至标线，摇匀。每毫升标准使用液含 1.000μg 六价铬。使用当天配制。

(9) 200g/L 尿素溶液。

(10) 20g/L 亚硝酸钠溶液。

(11) 二苯碳酰二肼溶液。称取二苯碳酰二肼（简称 DPC，$C_{13}H_{14}N_4O$）0.2g，溶于 50mL 丙酮中，加水稀释至 100mL，摇匀，贮于棕色瓶内，置于冰箱中保存。颜色变深后不能再用。

(12) 硝酸。

(13) 硫酸（$\rho=1.84g/mL$）。

(14) 三氯甲烷。

(15) （1+1）氨水。

(16) 50g/L 铜铁试剂。称取 5g 铜铁试剂 [$C_6H_5N(NO)ONH_4$]，溶于冰水中并稀释至 100mL。现用现配。

四、实验步骤

（一）六价铬的测定

1. 水样预处理

(1) 对不含悬浮物、低色度的清洁地表水，可直接进行测定。

(2) 如果水样有色但不深，可进行色度校正。即另取一份试样，加入除显色剂以外的各种试剂，以 2mL 丙酮代替显色剂，用此溶液为测定试样溶液吸光度的参比溶液。

(3) 对浑浊、色度较深的水样，应加入氢氧化锌共沉淀剂，并进行过滤处理。

(4) 水样中存在次氯酸盐等氧化性物质时，干扰测定，可加入尿素和亚硝酸钠消除。

(5) 水样中存在低价铁、亚硫酸盐、硫化物等还原性物质时，可将 Cr(Ⅵ) 还原为 Cr(Ⅲ)。此时，调节水样 pH 值至 8，加入显色剂溶液，放置 5min 后再酸化显色，并以同样的方法绘制标准曲线。

2. 标准曲线的绘制

取 9 支 50mL 比色管，依次加入 0、0.20、0.50、1.00、2.00、4.00、6.00、8.00 和 10.00mL 铬标准使用液，用水稀释至标线，加入 0.5mL（1+1）硫酸和 0.5mL（1+1）磷酸，摇匀。加入 2mL 显色剂溶液，摇匀。5~10min 后，于 540nm 波长处，用 1cm 或 3cm 比色皿，以水为参比，测定吸光度并作空白校正。以吸光度为纵坐标，相应六价铬含量为横坐标，用 Microsoft Office Excel 绘制标准曲线，并求线性回归方程。

3. 水样的测定

取适量（含铬少于 50μg）无色透明或经预处理的水样于 50mL 比色管中，用水稀释至标线，测定方法同标准溶液。进行空白校正后根据所测吸光度从标准曲线上查得 Cr（Ⅵ）质量。

4. 计算

水样中铬质量浓度 ρ（mg/L）按式（4-2-9）计算

$$\rho(\mathrm{Cr}, \mathrm{mg/L}) = \frac{m}{V} \tag{4-2-9}$$

式中　m——从标准曲线上查得的 Cr(Ⅵ) 质量，μg；
　　　V——水样的体积，mL。

（二）总铬量的测定

1. 水样预处理

一般清洁地表水可直接用高锰酸钾氧化后测定。

对含大量有机物的水样，需进行消解处理。取 50mL 或适量（含铬少于 50μg）水样，置于 150mL 烧杯中，加入 5mL 硝酸和 3mL 硫酸，加热蒸发至冒白烟。如溶液仍有色，再加入 5mL 硝酸，重复上述操作，至溶液清澈，冷却。用水稀释至 10mL，用氨水（1+1）中和至 pH 值 1~2，移入 50mL 容量瓶中，用水稀释至标线，摇匀，备用。

如果水样中钼、钒、铁、铜等含量较大，先用铜铁试剂-三氯甲烷萃取除去，然后再进行消解处理。

2. 高锰酸钾氧化三价铬

取 50.0mL 或适量（铬质量少于 50μg）清洁水样或经预处理的水样（如不到 50.0mL，用水补充至 50.0mL）于 150mL 锥形瓶中，用（1+1）氨水和硫酸溶液调至中性，加入几粒玻璃珠，加入（1+1）硫酸和（1+1）磷酸各 0.5mL，摇匀。加入 40g/L 高锰酸钾溶液 2 滴，如紫色消褪，则继续滴加高锰酸钾溶液至溶液保持紫红色。加热煮沸至溶液剩约 20mL。冷却后，加入 1mL 200g/L 的尿素溶液，摇匀。用滴管加 20g/L 亚硝酸钠溶液，每加一滴充分摇匀，至紫色刚好消失。稍停片刻，待溶液内气泡逸尽，转移至 50mL 比色管中，稀释至标线，供测定。

其余步骤同六价铬的测定。

五、数据处理

（1）实验数据记录见表 4-2-6。

表 4-2-6　实验数据记录

	标准曲线绘制									水样测定
水样体积/mL	0	0.2	0.5	1.0	2.0	4.0	6.0	8.0	10.0	
吸光度										
Cr 质量/μg										

（2）绘制标准曲线。

（3）计算水样中 Cr 含量。

六、注意事项

（1）用于测定铬的玻璃器皿不能用重铬酸钾洗液洗涤。

（2）Cr(Ⅵ)与显色剂的显色反应一般控制酸度在 $0.05\sim0.3$ mol/L（$1/2H_2SO_4$）范围，以 0.2mol/L 时显色最好。显色前，水样应调至中性。显色温度和放置时间对显色有影响，在15℃时，5～15min颜色即可稳定。

（3）如测定清洁地面水样，显色剂可按以下方法配制：溶解 0.2g 二苯碳酰肼于 100mL 95％乙醇中，边搅拌边加入 400mL（1+9）硫酸。该溶液在冰箱中可存放一个月。用此显色剂，在显色时直接加入 2.5mL 即可，不必再加酸。但加入显色剂后，要立即摇匀，以免 Cr(Ⅵ) 被乙酸还原。

实验十四　混凝实验

分散在水中的胶体颗粒带有电荷，同时在布朗运动及其表面水化膜作用下，长期处于稳定分散状态，不能用自然沉淀法去除。向这种水中投加混凝剂后，可以使分散颗粒相互结合聚集增大，从水中分离出来。

由于各种原水有很大差别，混凝效果不尽相同。混凝剂的混凝效果不仅取决于混凝剂投加量，同时还取决于水的 pH、水流速度梯度等因素。

一、实验目的

（1）观察混凝现象及过程，了解混凝的净水机理及影响混凝的重要因素。

（2）掌握求得某水样最佳混凝条件（投药量、pH）的基本方法。

二、实验原理

水中粒径小的悬浮物以及胶体物质，由于微粒的布朗运动、胶体颗粒间的静电斥力和胶体的表面作用，致使水中这种浑浊状态稳定。化学混凝的处理对象主要是废水中的微小悬浮物和胶体物质。根据胶体的特性，在废水处理过程中通常采用投加电解质、不同电荷的胶体或高分子等方法破坏胶体的稳定性，然后通过沉淀分离，达到废水净化效果的目的。关于化学混凝的机理主要有以下四种解释。

（一）压缩双电层机理

当两个胶粒相互接近以至双电层发生重叠时，就产生静电斥力。加入的反离子与扩散层原有反离子之间的静电斥力将部分反离子挤压到吸附层中，从而使扩散层厚度减小。由于扩散层变薄，颗粒相撞时的距离减小，相互间的吸引力变大。颗粒间排斥力与吸引力的合力由斥力为主变为以引力为主，颗粒就能相互凝聚。

（二）吸附电中和机理

异号胶粒间相互吸引达到电中和而凝聚；大胶粒吸附许多小胶粒或异号离子，ζ 电位降低，吸引力使异号胶粒相互靠近发生凝聚。

（三）吸附架桥机理

吸附架桥作用是指链状高分子聚合物在静电引力、范德华力和氢键力等作用下，通过活性部位与胶粒和细微悬浮物等发生吸附桥连的现象。

（四）沉淀物网捕机理

当采用铝盐或铁盐等高价金属盐类作凝聚剂时，投加量很大会形成大量的金属氢氧化物沉淀，可以网捕、卷扫水中的胶粒，水中的胶粒以这些沉淀为核心产生沉淀。这基本上是一种机械作用。

向水中投加混凝剂后，①能降低颗粒间的排斥能峰，降低胶粒的ζ电位，实现胶粒脱稳；②同时也能发生高聚物式高分子混凝剂的吸附架桥作用；③通过网捕作用，达到颗粒的凝聚。

消除或降低胶体颗粒稳定因素的过程叫作脱稳。脱稳后的胶粒，在一定的水力条件下，能形成较大的絮凝体，俗称矾花。直径较大且较密实的矾花容易下沉。自投加混凝剂直至形成较大矾花的过程叫混凝。在混凝过程中，上述现象往往同时存在，只是在一定情况下以某种现象为主。

三、实验仪器与试剂

（1）智能型混凝试验搅拌仪1台。
（2）PHS-2型酸度计1台。
（3）HACH 2100N浊度仪1台。
（4）烧杯（200mL，7个）。
（5）移液管（1mL、2mL、5mL、10mL各1支）。
（6）洗耳球1个，配合移液管移药用。
（7）量筒（1000mL，1个，量原水体积）。
（8）混凝剂：硫酸铝[$Al_2(SO_4)_3$]、聚合硫酸铁（PFS）、聚合氯化铝（PAC）、聚合硫酸铁铝（PAFS）、聚丙烯酰胺（PAM）等，浓度为1%或10g/L。
（9）盐酸、氢氧化钠（浓度为10%）。
（10）实验用原水（取河水或用黏土和自来水配成水样20L，静沉6h，其上清液为实验用原水）。

四、实验步骤

混凝实验分为最佳投药量、最佳pH、最佳水流速度梯度三部分。在进行最佳投药量实验时，先选定一种搅拌速度变化方式和pH，求出最佳投药量，然后按照最佳投药量求出混凝最佳pH。

（一）最佳投药量实验步骤

（1）确定原水特征，即测定原水水样浑浊度、pH、温度。
（2）确定形成矾花所用的最小混凝剂量。方法是通过慢速搅拌（50r/min）烧杯中800mL原水，并每隔1min增加1mL混凝剂投加量，直至出现矾花为止。这时的混凝剂量作为形成矾花的最小投加量。
（3）用6个1000mL的烧杯，分别放入800mL原水，置于混凝试验搅拌仪平台上。
（4）确定实验时的混凝剂投加量。根据步骤（2）得出的形成矾花最小混凝剂投加量，取其1/4作为1号烧杯的混凝剂投加量，分别取其1/2、3/4、1倍、3/2倍、2倍作为2～6

号烧杯的混凝剂投加量。加药时，把混凝剂分别加到仪器上 1～6 号加药管中，这样可以保证同时加药。

（5）启动搅拌机，快速搅拌 30 s，转速约 300r/min；中速搅拌 6min，转速约 100r/min；慢速搅拌 6min，转速约 50r/min。如果用污水进行混凝实验，污水胶体颗粒比较脆弱，搅拌速度可适当放慢。

（6）关闭搅拌机，抬起搅拌桨，静置沉淀 10min，取出上清液放入烧杯内，立即用浊度仪测定浊度（每杯水样测定三次），记入表 4-2-7 中。

（二）最佳 pH 实验步骤

（1）用 6 个 1000mL 的烧杯，分别放入 800mL 原水，置于混凝试验搅拌仪平台上。

（2）调整原水 pH，用移液管依次向 1 号、2 号、3 号装有水样的烧杯中分别加入 1.5mL、1.0mL、0.5mL 10％浓度的盐酸。依次向 5 号、6 号装有水样的烧杯中分别加入 0.5mL、1.0mL 10％浓度的氢氧化钠。

（3）启动搅拌机，快速搅拌 30s，转速约 300r/min。用酸度计测定各水样的 pH，记入表 4-2-8 中。

（4）利用仪器的加药管，向各烧杯中加入相同剂量的混凝剂（最佳剂量采用最佳投药量实验中得出的最佳投药量结果）。

（5）启动搅拌机，快速搅拌 30s，转速约 300r/min；中速搅拌 6min，转速约 100r/min；慢速搅拌 6min，转速约 50r/min。如果用污水进行混凝实验，污水胶体颗粒比较脆弱，搅拌速度可适当放慢。

（6）关闭搅拌机，抬起搅拌桨，静置沉淀 10min，取出上清液放入烧杯内，立即用浊度仪测定浊度（每杯水样测定三次），记入表 4-2-8 中。

五、数据处理

混凝剂_____　　混凝剂浓度_____
原水浊度_____　　原水 pH _____　　原水温度_____
最小混凝剂量（mL）_____　　相当于（mg/L）_____

（一）最佳投药量实验结果整理

（1）把原水特征、混凝剂投加情况、沉淀后的剩余浊度记入表 4-2-7 中。

表 4-2-7　最佳混凝剂投加量

水样编号		1	2	3	4	5	6
投药量/mL							
出现矾花时间/min							
矾花沉淀情况(快/慢)							
剩余浊度/NTU	1						
	2						
	3						

（2）以剩余浊度为纵坐标，投药量为横坐标，绘制剩余浊度与投药量关系曲线，从曲线上可求得不大于某一剩余浊度的最佳投药量值。

（二）最佳 pH 实验结果整理

（1）把原水特征、混凝剂投加情况、酸碱加注情况、沉淀后的剩余浊度记入表 4-2-8 中。

表 4-2-8　最佳 pH（投药量：_____ mL）

水样编号		1	2	3	4	5	6
HCl 体积/mL							
NaOH 体积/mL							
水样 pH							
剩余浊度/NTU	1						
	2						
	3						

（2）以剩余浊度为纵坐标，水样 pH 为横坐标，绘制剩余浊度与 pH 关系曲线，从图上求出所投加混凝剂的混凝最佳 pH 及其适用范围。

六、思考题

（1）根据实验结果以及实验中所观察到的现象，简述影响混凝的几个主要因素。
（2）为什么最大投药量时，混凝效果不一定好？
（3）根据最佳投药量实验曲线，分析沉淀水浊度与混凝剂投加量的关系。

七、注意事项

（1）在最佳投药量、最佳 pH 实验中，向各烧杯投加药剂时要求同时投加，避免因时间间隔较长各水样加药后反应时间长短相差太大，混凝效果悬殊。
（2）在测定水的浊度、抽吸上清液时，不要扰动底部沉淀物。同时，各烧杯抽吸的时间间隔尽量减小。

实验十五　非色散红外吸收法测定水中总有机碳含量

总有机碳是指水体中溶解性和悬浮性有机物含碳的总量，常以 TOC 表示。水中有机物的种类很多，除含碳外，还含有氢、氮、硫等元素，不能全部进行分离鉴定。TOC 是一个快速检定的综合指标，它以碳的数量表示水中含有机物的总量。但由于它不能反映水中有机物的种类和组成，因而不能反映总量相同的总有机碳所造成的不同污染后果。由于 TOC 的测定采用燃烧法，因此能将有机物全部氧化，它比 BOD_5 或 COD 更能直接表示有机物的总量，通常作为评价水体有机物污染程度的重要依据。TOC 分析已成为世界许多国家水处理和质量控制的主要项目。

总有机碳（TOC）是以碳的含量表示水中有机物的总量，结果以碳（C）的质量浓度（mg/L）表示。碳是一切有机物的共同成分，也是组成有机物的主要元素，水的 TOC 值越高，说明水中有机物含量越高，因此，TOC 可以作为评价水质有机污染的指标。当然，它排除了其他元素，如含高 N、S 或 P 等元素的有机物在燃烧氧化过程中，同样参与了氧化反应，但 TOC 以 C 计结果中并不能反映出这部分有机物的含量。

TOC 的测定采用仪器法，按工作原理不同，可分为燃烧氧化-非色散红外吸收法、电导法、湿法氧化-非色散红外吸收法等。其中燃烧氧化-非色散红外吸收法流程简单、重现性好、灵敏度高，在国内外被广泛采用。燃烧氧化-非色散红外吸收法测定 TOC 又分为差减法和直接法两种。由于个别含碳有机物在高温下不易被燃烧氧化，因此所测得的 TOC 值常稍低于理论值。

TOC 测定仪具有流程简单、重现性好、灵敏度高、稳定可靠、测定过程一般不消耗化学药品、不产生二次污染、可测量全部有机碳含量等优点，因而 TOC 是监测有机污染物排放的最佳综合指标。

因此，掌握非色散红外吸收法测定水中总有机碳含量具有十分重要的意义。

一、实验目的

(1) 掌握非色散红外吸收法测定水中总有机碳含量的基本原理和操作步骤。
(2) 了解总有机碳含量对水质评价的影响。

二、实验原理

（一）差减法测定总有机碳

将试样连同净化空气（干燥并除去二氧化碳）分别导入高温燃烧管（900℃）和低温反应管（160℃）中，经高温燃烧管的水样受高温催化氧化，使有机化合物和无机碳酸盐均转化成为二氧化碳，经低温反应管的水样酸化使无机碳酸盐分解成二氧化碳，其所生成的二氧化碳依次被引入非色散红外线检测器。由于一定波长的红外线被二氧化碳选择吸收，在一定浓度范围内二氧化碳对红外线吸收的强度与二氧化碳的浓度成正比，故可对水样总碳（TC）和无机碳（IC）进行定量测定。

（二）直接法测定总有机碳

将水样酸化后曝气，将无机碳酸盐分解生成二氧化碳驱除，再注入高温燃烧管中，可直接测定总有机碳。

三、实验仪器与试剂

（一）仪器

(1) 非色散红外吸收 TOC 分析仪。工作条件：环境温度为 5～35℃；工作电压为仪器额定电压，交流电；总碳燃烧管温度选定 900℃；无机碳反应管温度控制在 (160±5)℃；载气流量为 180mL/min；50.00μL 微量注射器；10mL 具塞比色管。
(2) 一般实验室仪器。

（二）试剂

除另有说明外，均为分析纯试剂，所用水均为无二氧化碳蒸馏水。

(1) 无二氧化碳蒸馏水：将重蒸馏水在烧杯中煮沸蒸发（蒸发量10%），稍冷，装入插有碱石灰管的下口瓶中备用。
(2) 邻苯二甲酸氢钾（$KHC_8H_4O_4$）：优级纯。
(3) 无水碳酸钠（Na_2CO_3）：优级纯。
(4) 碳酸氢钠（$NaHCO_3$）：优级纯，存放于干燥器中。

（5）有机碳标准贮备溶液（400mg/L）。称取邻苯二甲酸氢钾（预先在110~120℃干燥2h，置于干燥器中冷却至室温）0.8500g，溶解于水中，移入1000mL容量瓶内，用水稀释至标线，混匀，在低温（4℃）冷藏条件下可保存48d。

（6）有机碳标准溶液（80mg/L）。准确移取10.00mL有机碳标准贮备溶液，置于50mL容量瓶内，用水稀释至标线混匀，此溶液用时现配。

（7）无机碳标准贮备溶液（400mg/L）。称取碳酸氢钠（预先在干燥器中干燥）1.400g和无水碳酸钠（预先在105℃干燥2h，置于干燥器中，冷却至室温）1.770g，溶解于水中，转入1000mL容量瓶内，用水稀释至标线，混匀。

（8）无机碳标准溶液（80mg/L）。准确吸取10.00mL无机碳标准贮备溶液（7），置于50mL容量瓶中，用水稀释至标线，混匀。此溶液现时现配。

四、实验步骤

（一）样品采集与保存

水样采集后，必须贮存于棕色玻璃瓶中。常温下水样可保存24h，如不能及时分析，水样可加硫酸将其pH调至≤2，于4℃冷藏，可保存7d。

（二）仪器的调试

按说明书调试TOC分析仪及记录仪；选择好灵敏度、测量范围档、总碳燃烧管温度及载气流量，仪器通电预热2h，至红外线分析仪输出时记录仪上的基线趋于稳定。

（三）干扰的排除

水样中常见共存离子含量超过干扰允许值时，会影响红外线的吸收。这种情况下，必须用无二氧化碳蒸馏水稀释水样，至各共存离子含量低于其干扰允许浓度后，再行分析。

（四）差减测定法

经酸化的水样，在测定前应用氢氧化钠溶液中和至中性，用$50.0\mu L$微量注射器分别准确吸取混匀的水样$20.0\mu L$，依次注入总碳燃烧管和无机碳反应管，测定记录仪上出现的相应的吸收峰峰高。

（五）直接测定法

将已酸化的约25mL水样移入50mL烧杯中，在磁力搅拌器上剧烈搅拌几分钟或向烧杯中通入无二氧化碳的氮气，以除去无机碳。吸取$20.0\mu L$除去无机碳的水样，注入总碳燃烧管，测量记录仪上出现的吸收峰峰高。

（六）空白试验

按上述步骤进行空白试验，用$20.0\mu L$水代替试样。

（七）校准曲线的绘制

在一组七个10mL具塞比色管中，分别加入0.00、0.50、1.50、3.00、4.50、6.00及7.50mL有机碳标准溶液、无机碳标准溶液，用蒸馏水稀释至标线，混匀。配制成0.0、4.0、12.0、24.0、36.0、48.0及60.0mg/L的有机碳和无机碳标准系列溶液。然后按上述步骤操作，用测得的标准系列溶液吸收峰峰高，减去空白试样吸收峰峰高，得校正吸收峰峰高，由标准系列溶液浓度与对应的校正吸收峰峰高绘制有机碳和无机碳校准曲线。亦可按线性回归方程的方法，计算出校准曲线的直线回归方程。

五、数据处理

（一）差减测定法

根据所测试样吸收峰峰高减去空白试样吸收峰峰高的校正值，从校准曲线上查得或由校准曲线回归方程算得总碳（TC，mg/L）和无机碳（IC，mg/L）的质量浓度，总碳与无机碳质量浓度之差值，即样品总有机碳（TOC，mg/L）的质量浓度：

$$TOC（mg/L）=TC（mg/L）-IC（mg/L）$$

（二）直接测定法

根据所测试样吸收峰峰高减去空白试样吸收峰峰高的校正值，从校准曲线上查得或由校准曲线回归方程算得总碳（TC，mg/L）的质量浓度，即样品总有机碳（TOC，mg/L）的质量浓度：

$$TOC（mg/L）=TC（mg/L）$$

进样体积为 20.0 μL，其结果以一位小数表示。

六、思考题

（1）试分析讨论水中总有机碳的主要来源。

（2）哪些因素会干扰水中总有机碳的测定，如何消除？

第三节　土壤环境化学实验

实验十六　石墨炉原子吸收光谱法测定土壤中的铅

一、实验目的

（1）理解石墨炉原子吸收光谱法的原理及特点。

（2）掌握石墨炉原子吸收光谱仪的操作技术。

（3）熟悉土壤样品的前处理。

二、实验原理

铅是一种具有积蓄性的有害元素，当摄入过多时，会对神经系统、消化系统和造血系统造成危害。采用盐酸-硝酸-氢氟酸-高氯酸全消解的方法，彻底破坏土壤的矿物晶格，使试样中的待测元素全部进入试液。将试液注入石墨炉中，经过预先设定的干燥、灰化、原子化等程序使共存基体成分蒸发除去，同时在原子化阶段的高温下铅化合物离解为基态原子蒸气，并对空心阴极灯发射的特征谱线产生选择性吸收。在选择的最佳测定条件下，通过背景扣除，测定试液中铅的吸光度。

三、实验仪器与试剂

（一）仪器

石墨炉原子吸收分光光度计（带有背景扣除装置）、铅空心阴极灯、氩气钢瓶。

（二）试剂

(1) 浓盐酸、浓硝酸、氢氟酸、高氯酸，均为优级纯。
(2) 硝酸溶液，1+5 配制。
(3) 磷酸氢二铵（优级纯）水溶液，质量分数为 5％。
(4) 土壤试样。
(5) 铅标准储备液，0.500mg/mL。准确称取 0.500g 光谱纯金属铅于 50mL 烧杯中，加入 20mL 硝酸溶液，微热溶解，冷却后转移至 1000mL 容量瓶中，用水定容至标线，摇匀。
(6) 铅标准使用液，250μg/L。临用前将铅标准储备液用硝酸溶液逐级稀释配制得到。

四、实验步骤

(1) 准确称取经预处理后的土壤样品 0.1～0.3g（精确到 0.0002g）于 50mL 聚四氟乙烯坩埚中，用水润湿后加入 5mL 浓盐酸，于通风橱内的电热板上低温加热使样品初步分解，当试液蒸发至约 2～3mL 时，取下稍冷，加入 5mL 浓硝酸、2mL 氢氟酸，2mL 高氯酸，加盖后于电热板中温加热 1h 左右，开盖，继续加热除去硅，为达到良好的去硅效果，应经常摇动坩埚。当加热至冒浓厚高氯酸白烟时，加盖，使黑色有机碳化物充分分解。待坩埚上的黑色有机物消失后，开盖驱赶白烟并蒸至内容物呈黏稠状。视消解情况，可再加入 2mL 浓盐酸、2mL 氢氟酸，1mL 高氯酸，重复上述消解过程。当白烟再次基本冒尽且内容物呈黏稠状时，取下稍冷，用水冲洗坩埚盖和内壁，并加入 1mL 硝酸溶液温热溶解残渣。再将溶液转移至 25mL 容量瓶中，加入 3mL 磷酸氢二铵溶液，冷却后定容，摇匀备测。

由于土壤种类多，有机质含量差异较大，消解时各种酸的用量可视消解情况酌情增减。土壤消解液应呈白色或淡黄色，没有沉淀物存在。

(2) 按照仪器说明书调节仪器至最佳工作条件，测定试液的激光度。仪器测量条件参考表 4-3-1。

表 4-3-1 仪器测量条件

测定波长/nm	283.3	原子化/(℃/s)	2000/5
通带宽度/nm	1.3	清除/(℃/s)	2700/3
灯电流/mA	7.5	氩气流量/(mL/min)	200
干燥/(℃/s)	80～100/20	原子化阶段是否停气	是
灰化/(℃/s)	700/20	进样量/μL	10

(3) 空白实验。用水代替试样，进行步骤 (1)、(2)。
(4) 校准曲线。用移液管准确移取铅标准使用液 0.00mL、0.50mL、1.00mL、2.00mL、3.00mL、5.00mL 于 25mL 容量瓶中。加入 3.0mL 磷酸氢二铵溶液，用硝酸溶液定容。该标准溶液含铅 0.0μg/L、5.0μg/L、10.0μg/L、20.0μg/L、30.0μg/L、50.0μg/L。按照步骤 (2) 中条件由低到高浓度依次测定标准溶液的吸光度。用减去空白的吸光度与对应的元素含量（μg/L）绘制铅标准曲线。
(5) 根据扣除空白吸光度后的样品吸光度，在标准曲线上查出待测样品中的铅浓度。

五、数据处理

土壤样品中铅的质量分数 W（μg/kg）按式（4-3-1）计算：

$$W = \frac{cV}{m(1-f)} \tag{4-3-1}$$

式中　W——土壤样品中铅的质量分数，$\mu g/kg$；

　　　c——试液的吸光度减去空白试样的吸光度，然后在标准曲线上查得铅的质量浓度（$\mu g/L$）；

　　　V——试样溶液的体积，mL；

　　　m——称取试样的质量，g；

　　　f——试样中水分的含量，%。

土壤水分含量 f 的测定。称取通过100目筛的风干土样 5～10g（精确至0.01g），置于铅盒或称量瓶中，在105℃烘箱中烘 4～5h，烘干至恒重。以百分数表示的风干土样水分含量 f(%) 按式（4-3-2）计算。

$$f = \frac{W_1 - W_2}{W_1} \tag{4-3-2}$$

式中　f——土样中水分的含量，%；

　　　W_1——烘干前土样质量，g；

　　　W_2——烘干后土样质量，g。

六、思考题

（1）石墨炉原子吸收分光光度计主要包括哪几部分，各部分的作用是什么？

（2）实验中测定空白溶液的吸光度有何意义？

（3）石墨炉原子吸收光谱法与火焰原子吸收光谱法相比，其优缺点是什么？

七、注意事项

（1）消解过程中使用的氢氟酸、高氯酸等有爆炸危险，整个消解过程一定要在通风橱中进行。

（2）电热板的温度不宜太高，否则会使聚四氟乙烯坩埚变形。

实验十七　重金属污染土壤的化学修复（EDTA对土壤中铜的淋洗）

土壤污染对人类的危害性极大，它不仅会直接导致土地质量退化，影响农产品安全，而且会通过食物链对人体健康产生影响，还会通过对地下水的污染及污染的转移造成对人类生存环境多个层面上的危害。鉴于土壤污染危害的严重性，污染土壤的修复在国际上受到了高度重视，并成为国内外环境研究热点。

一、实验目的

（1）掌握土壤预处理的一般方法。

（2）学习淋洗法修复重金属污染土壤的原理。

（3）探讨EDTA修复重金属污染土壤的最优条件，评价修复效果。

二、实验原理

土壤淋洗法是土壤修复方法的一种，该方法是一个污染土壤和淋洗液间高能量接触，从污泥、沉积物中去除有机和无机污染物的过程，包括了物理、化学多机制的修复工艺，能够实现危险物质的分离、隔离或无害化转变，然后对含有污染物的淋洗液进行处理，能够减少异位修复采集、运送、复原的费用，并且在修复有机、无机污染方面均有很大潜力，在经济上与其他方法相比也有一定的优势，因此在国际上受到广泛关注。淋洗法修复重金属污染土壤，通常采用在淋洗液中添加无机酸、低相对分子质量有机酸或合适的络合剂如 EDTA、NTA 等来增加金属的可移动性，从而达到清洁土壤的目的。

乙二胺四乙酸钠的结构为：

$$\begin{array}{c}NaOOCH_2C \\ NaOOCH_2C\end{array} \!\!\!\! > \!\! N\!-\!H_2C\!-\!CH_2\!-\!N \!\! <\!\!\! \begin{array}{c}CH_2COOH \\ CH_2COONa\end{array}$$

它有 4 个给出电子对的氧原子和 2 个给出电子对的氮原子，故可和绝大多数金属离子形成稳定的螯合物。Cu^{2+} 与乙二胺四乙酸（EDTA）的作用方程式为：

$$Cu^{2+} + Y^{4-} \rightleftharpoons CuY^{2-}$$

Cu^{2+} 和 Y^{4-} 螯合形成五个原子环，这是螯合物特别稳定的主要原因。

用 EDTA 淋洗土壤中的 Cu 相当于 EDTA 对土壤中各种键合形态的 Cu 的竞争平衡作用。

三、实验仪器与试剂

（一）仪器

塑料离心管、移液管、移液枪、摇床、容量瓶、离心机、酸度计、原子吸收测定仪。

（二）试剂

(1) $CuCl_2$ 标准溶液：准确称取 26.85mg 的 $CuCl_2 \cdot H_2O$，加入 1～2mL 稀盐酸（0.1mol/L）溶解，防止 Cu^{2+} 水解，再将溶液转移至 1L 容量瓶中，用去离子水定容至刻度线，摇匀。

(2) EDTA 水溶液：称取 18.61g EDTA 二钠盐，溶于 500mL 去离子水中，用 1mol/L NaOH 调节 pH 至 5.0，将溶液转移至 1L 容量瓶中，用去离子水定容至刻度线，摇匀。

(3) 0.05mol/L $NaNO_3$ 溶液：准确称取 4.25g $NaNO_3$，转移至烧杯，加入 500mL 去离子水，搅拌至完全溶解，将溶液转移至 1L 容量瓶中，用去离子水定容至刻度线，摇匀。

(4) HCl：优级纯。

(5) HNO_3：优级纯。

(6) $HClO_4$：优级纯。

(7) NaOH：优级纯。

四、实验步骤

（一）污染土壤的制备

把野外采回的土样倒在塑料薄膜或纸上，趁半干状态把土块压碎，除去残根、石块等杂物，铺成薄层，在阴凉处慢慢晾干。风干后土样用硬棒碾碎后，过 2mm 尼龙筛，对于大于 2mm 砂砾应计算其占整个土样的百分数。将小于 2mm 的土样，反复用四分法取样，样品进一步用玛瑙研钵研细，过 100 目筛，整瓶待用。

（二）污染土样 Cu 含量的确定

1. 土壤样品的消化

准确称取 1.000g 土样（3 份）及土壤标样（3 份）于 100mL 烧杯中，用少量去离子水润湿，缓慢加入 5mL 王水，盖上表面皿。同时做一份试剂空白，把烧杯放在通风橱内的电炉上加热，开始低温，慢慢提高温度，并保持微沸状态，使其充分分解。注意消化温度不宜过高，防止样品外溅，操作时应戴上手套和护目镜。当激烈反应完毕，大部分有机物分解后，取下烧杯冷却，沿烧杯壁加入约 3mL 高氯酸，继续加热分解至冒白烟，样品变为灰白色，移去表面皿，赶出过量的高氯酸，把样品蒸至近干，取下冷却，加入 5mL 1% 稀硝酸加热，冷却后用中速定量滤纸过滤到 50mL 容量瓶中，滤渣用 1% 的硝酸洗涤，最后定容，摇匀待测。

2. 配备标准曲线溶液

取 6 个 50mL 容量瓶，分别加入 5 滴 1：1 盐酸，依次加入 0.00mL、2.00mL、4.00mL、6.00mL、8.00mL、10.00mL 质量浓度为 10mg/L 的铜标准液，用去离子水稀释至刻度，摇匀。

3. 测定

将标准溶液和消化液直接喷入空气-乙炔火焰中，用原子吸收检测仪测定吸收值。

（三）确定洗脱率与pH的关系

取 15 个 50mL 塑料离心管，分别称取 0.5000g 铜污染土样。

在 1000mL 烧杯中加入 600.0mL 去离子水，加入适量的 EDTA，调节体系中 EDTA 的物质的量与 Cu 的物质的量比为 4：1（Cu 含量根据配备标准曲线溶液中确定的量进行计算），假设测得污染土样中 Cu 的质量分数为 c，每个离心管中将加入淋洗剂 30.0mL，则淋洗体系中 Cu 的物质的量浓度按式（4-3-3）计算。

$$\text{Cu 物质的量浓度（mol/L）} = \frac{0.500c}{63.5 \times 0.030} \quad (4\text{-}3\text{-}3)$$

用硝酸钠调节离子强度，使 EDTA 水溶液中的硝酸钠浓度为 0.01mol/L，再用 HCl 和 NaOH 调节溶液的 pH，先把溶液 pH 调为 4，分别吸取 30mL 至 3 个离心管中，然后再把溶液 pH 调至 5，分别吸取 30mL 至 3 个离心管中，以此类推，分别作 4 个 pH 点（4、5、6、7）。然后把加好离心剂的离心管置于摇床中振摇，250 次/min，振摇 24h 后取出。于 2400r/min 下离心 15min，取上层清液 5mL 进行原子吸收光谱（AAS）测定，剩余溶液测 pH。

（四）测定洗脱率与物质的量比值的关系

取 15 个 50mL 的塑料离心管，分别称取 0.5000g 铜污染土样置于其中。

取 5 个 250mL 烧杯，分别加 150mL 的去离子水，然后分别加入适量的 EDTA，使得各烧杯中 EDTA 与 Cu 的物质的量比分别为 2：1，4：1，6：1，8：1，10：1，Cu 和 EDTA 的量的确定方法同步骤（三），用硝酸钠调节溶液的离子强度，使各 EDTA 水溶液中的硝酸钠浓度为 0.01mol/L。

根据步骤（三）确定的最优 pH，分别调节 5 个烧杯中 EDTA 水溶液淋洗剂的 pH，调好 pH 的淋洗剂，各取 3 份 30mL 于 3 个称好铜污染土样的离心管中。离心管置于摇床中，250 次/min 条件下振摇 24h，取出，于 4500r/min 下离心 15min，取上层清液 5mL 进行 AAS 测定，剩余溶液测定 pH。

（五）确定洗脱率和时间的关系

基本步骤同上，根据以上确定的条件，调节物质的量比、离子强度和 pH，一共做 13 个点，每个点 3 份平行，0h 到 24h，每隔两小时取一次样，离心，取上层清液 5mL AAS 测定，剩余溶液测 pH。

五、数据处理

（一）污染土样铜的含量

1. 标准曲线的绘制

测得的标准溶液的吸收值对浓度（mol/L）作图。

2. 计算土样铜含量

根据所测得的吸收值（如试剂空白有吸收，则应扣除空白吸收值）在标准曲线上得到的相应的质量浓度 M（mg/mL），则土样中 Cu 的质量分数按式（4-3-4）计算。

$$\text{Cu 质量分数} = \frac{MV}{m} \times 10^{-3} \tag{4-3-4}$$

式中 M——标准曲线上得到的相应的质量浓度，mg/mL；

　　　V——定容体积，mL；

　　　m——土样质量，g。

（二）洗脱率与 pH 的关系

1. 标准曲线的绘制

测得的标准溶液的吸收值对浓度（mol/L）作图。

2. 绘制洗脱率与 pH 关系图

淋洗体系（离心管中 30mL 溶液）中铜的物质的量浓度按式（4-3-3）计算：

所测得的吸收值在标准曲线得到相应的浓度 C_f（mol/L），则 Cu 的洗脱率按式（4-3-5）计算：

$$\text{洗脱率} = \frac{C_f}{C} \tag{4-3-5}$$

根据计算出的各 pH 点的洗脱率对 pH 作图，找出洗脱率与 pH 的关系。

（三）洗脱率与物质的量比值的关系

1. 标准曲线的绘制

测得的标准溶液的吸收值对浓度（mol/L）作图。

2. 绘制洗脱率与物质的量比值的关系图

C、C_f 及洗脱率的计算同上。根据计算出的各物质的量比值处的洗脱率对物质的量比值作图，找出洗脱率与物质的量比值的关系。

（四）洗脱率与时间的关系

测得的标准溶液的吸收值对浓度（mol/L）作图。

C、C_f 及洗脱率的计算同上。作洗脱率与时间的关系图，找出洗脱率与时间的关系。

六、思考题

（1）土壤淋洗过后，为什么还要测一次 pH？

（2）在进行土壤原位修复时，如何回收淋洗液？

（3）试讨论如何重复使用淋洗剂。

七、注意事项

（1）化学方法治理和修复重金属污染土壤就是利用化学试剂、化学反应或化学原理来降低土壤中重金属的迁移性和生物可利用率，减少甚至清除土壤中的重金属，从而达到污染土壤的治理和修复的目的。它包括淋洗法、固化法、施用改良剂法和电化学法等。

① 淋洗法就是用淋洗液淋洗被污染的土壤，又称洗土法或萃取法。在操作上又分为就地淋洗和移土清洗。在大多数情况下，为提高洗脱效率，需要在淋洗液中添加化学助剂，通常选择价格低廉、可生物降解、不易造成土壤污染的化学物质，最常用的是酸和螯合剂。后者更适用于对土壤的就地淋洗，因为它们对环境的危害更小，EDTA是对重金属最有效的提取剂。

② 固化处理是采用物理方法使土壤中的重金属固定化，使其危害降低。由于水泥廉价、易得，以水泥为黏合剂的稳定/固化技术已在国外用于无机物、有机物、核废物的最终处理技术，该技术不适用于污染耕地的就地修复。

③ 施用改良剂法是通过加入特定的化学物质调节土壤环境，控制反应条件，改变污染物的形态、水溶性、迁移性和生物有效性，使污染物钝化，降低其对生态环境的危害，包括沉淀作用、化学还原法、吸附法、拮抗作用、有机质改良等。

④ 电化学修复是一项正在发展的去除土壤重金属和放射性元素的就地修复技术。其原理为在一定的电流和电压作用下，不同离子能在电池和电迁移的作用下向相反的电极迁移，氢离子和金属离子向阴极定向移动，同时溶解土壤中的金属离子。此法受土壤性质的限制，不适用于渗透性较高、传导性较差的土壤及砂性土壤。

(2) 在消化土壤的过程中，应细心控制温度（升温过快反应物易溢出或炭化），待土壤里大部分有机质消化完冷却后再添加高氯酸。

(3) 本实验主要探讨了化学方面的机制，没有过多地涉及动力学过程，在研究土壤原位修复的时候，可以先通过原装土柱实验来摸索动力学条件。

(4) 淋洗剂的后续处理和循环利用是原位修复非常关键的部分，处理好这个问题对修复的效果和成本都很有帮助。

(5) 植物可以从土壤中吸收铜，但作物中铜的累积与土壤中总铜含量无明显相关，而与有效态铜（能被植物直接吸收利用的水溶态铜和交换态铜）含量密切相关。

实验十八 土壤有机质的测定

土壤有机质是指土壤中含碳的有机化合物。土壤中有机质的来源十分广泛：植物残体、动物残体、微生物残体、排泄物和分泌物、废水废渣。土壤有机质的含量在不同土壤中差异很大，含量高的可达20%或30%以上（如泥炭土、某些肥沃的森林土壤等），含量低的不足1%或0.5%（如荒漠土和风沙土等）。在土壤学中，一般把耕作层中含有机质20%以上的土壤称为有机质土壤，含有机质在20%以下的土壤称为矿质土壤。一般情况下，耕作层土壤有机质含量通常在5%以上。

土壤有机质的测定有如下办法：

(1) 目视比色法。通过以葡萄糖溶液为标准物质作参比，用重铬酸钾溶液氧化土壤有机质，氧化后的溶液颜色与有机质含量呈直线相关，可直接目视比得出结果。

(2) 灼烧法。测定土壤有机质中的碳经灼烧后造成的土壤失重：将在105℃下除去吸湿水的土样称重后于350~1000℃灼烧2h，再称重。两次称重之差即土壤中有机碳的质量。

(3) 光度法。以硫酸亚铁溶液为标准溶液进行土壤有机质的分光光度测定。

（4）重铬酸钾容量法——直接加热消解法。在加热条件下，以过量的 $K_2CrO_7 \cdot H_2SO_4$ 溶液氧化土壤中的有机质，以 $FeSO_4$ 标准溶液滴定剩余的 K_2CrO_7 来确定土壤中有机质的含量。

目视比色法简单易行，能快速得出分析结果，但测定的结果误差较大。灼烧法测定方法简单，但其灼烧时间较长，灼烧温度高，测定的结果误差较大，有一定的局限性。但这两种方法测定方便快捷、污染小。直接加热消解法针对油浴加热在加热条件上做出了改进，光度法是将重铬酸钾容量法的滴定改为光度测定，此两种方法测定结果具有很好的数据相关性和准确性，但若遇到大批量的样品时，直接加热消解法费时，而光度法却能在保证测定结果准确性的同时做到大批量快速测定。

土壤有机质是土壤中各种营养元素特别是氮的重要来源。一般来说，土壤有机质含量的多少是土壤肥力高低的一个重要指标。

一、实验目的

掌握重铬酸钾容量法测定土壤有机质。

二、实验原理

土壤有机质的测定中使用的比较普遍是的重铬酸钾容量法。虽然各种容量法所用的氧化剂及其浓度或具体条件稍有差异，但其基本原理都是在硫酸存在的条件下，用过量的重铬酸钾（$K_2Cr_2O_7$）溶液氧化土壤有机碳，再用硫酸亚铁溶液滴定剩余的重铬酸钾，根据消耗的重铬酸钾的量来间接计算土壤中有机碳的含量，进而根据土壤中有机质与有机碳的比例（即换算因数）计算土壤中有机质的含量。

采用 $K_2Cr_2O_7$ 并不能完全氧化土壤中的有机化合物，因此需要用一个校正系数来矫正未反应的有机碳的含量，一般认为该方法所氧化的有机碳仅为实际含量的 90%，即校正系数为 1.1，该方法具体反应过程如下：

氧化反应：$2K_2Cr_2O_7+8H_2SO_4+3C \longrightarrow 2K_2SO_4+2Cr_2(SO_4)_3+3CO_2+8H_2O$

滴定反应：$K_2Cr_2O_7+6FeSO_4+7H_2SO_4 \longrightarrow K_2SO_4+Cr_2(SO_4)_3+3Fe_2(SO_4)_3+7H_2O$

在滴定的过程中，使用邻菲罗啉氧化还原指示剂来指示滴定终点。在整个反应过程中溶液颜色的变化为：滴定开始以重铬酸钾的橙色为主，滴定过程中出现了 Cr^{3+} 的绿色，由于同时生成了三价铁，使得溶液整体呈现蓝绿色，当过量的重铬酸钾强氧化剂消耗完毕，标准硫酸亚铁溶液过量半滴，二价铁离子会与邻菲罗啉指示剂结合呈现棕红色，表示已到滴定终点。

三、实验试剂与仪器

（一）仪器

电子天平、玻璃试管、三角瓶、小漏斗、5mL 移液管 2 支、烘箱、360℃温度计 1 支、玻璃棒 1 支、25mL 活塞滴定管 1 支。

（二）试剂

（1）二氧化硅：粉末状。

（2）邻菲罗啉指示剂：称取邻菲罗啉 1.485g 溶于含有 0.695g 硫酸亚铁的 100mL 蒸馏水中，此指示剂易变质，应密封保存于棕色瓶中备用。

（3）0.068mol/L $K_2Cr_2O_7$ 溶液：称分析纯重铬酸钾 20g 溶于 500mL 蒸馏水中，冷却后定容至 500mL，装入棕色试剂瓶中。

（4）0.1mol/L $FeSO_4$ 溶液：称取化学纯硫酸亚铁 28g，溶于 600～800mL 水中，加浓硫酸 20mL，搅拌均匀，加水定容到 1L，储存于棕色瓶中，使用时必须每天用物质的量浓度为 0.068mol/L 重铬酸钾滴定。

（5）0.068mol/L 重铬酸钾标准溶液：取一定量的样品，130℃烘干 1.5h 后，称量 2g 溶于少量水，最后定容于 100mL 容量瓶中，放于瓶中保存。

四、实验步骤

（一）试样

选取有代表性的风干土壤样品，用镊子挑除植物根叶等有机残体，把土块压细，指通过 1mm 筛，充分混合后磨细，并全部通过 0.25mm 筛，装入磨口瓶中备用。用减量法称取约 0.1～0.5g 土壤样品于称量管中，质量记为 m；并将土样从称量管移入干燥的三角瓶中。

（二）加入氧化剂

用移液管吸取 0.068mol/L 重铬酸钾溶液 5mL、浓硫酸 5mL 加入有样品的三角瓶中，在每个三角瓶口插一个小玻璃漏斗。

（三）烘箱加热

将恒温箱的温度升至 185℃，预热好后，放入待测样品加热，让溶液在 170～185℃条件下沸腾 5min。

注意：三角瓶内起初有少量的小气泡产生，但并不是沸腾，而是有机质被氧化而释放出的二氧化碳。尤其是含碳酸盐的土壤，刚加热就有大量的二氧化碳产生，必须在真正沸腾时才能开始计时。

（四）滴定

取出烘箱中的三角瓶，待其冷却后用蒸馏水冲洗小漏斗和三角瓶内壁，待测液的总体积应控制在 30～35mL，加入 3 滴邻菲罗啉指示剂。用 0.1mol/L $FeSO_4$ 溶液滴定，记录滴定体积 V_2。

（五）空白

不加土壤样品，其他步骤与土壤测定相同，但滴定前溶液总体积控制在 20～25mL 为宜。通过空白实验可以求出滴定 5mL 浓度为 0.068mol/L 重铬酸钾及 5mL 浓硫酸溶液所需要的 0.1mol/L 硫酸亚铁溶液的用量 V_1。

五、数据处理

$$\text{有机碳质量分数}(\%) = \frac{(V_1-V_2)c \times \frac{1}{6} \times \frac{3}{2} \times 12 \times 1.1}{m \times 10^3} \times 100$$

式中　V_1——空白滴定用去 $FeSO_4$ 体积，mL；

　　　V_2——样品滴定用去 $FeSO_4$ 体积，mL；

c——标定的 $FeSO_4$ 溶液的浓度，mol/L；
　　m——风干土样质量，g；
　　12——碳原子的摩尔质量，g/mol；
　1.1——校正系数。

六、思考题

　　灼烧法测定土壤有机质含量是如何操作的？其适用范围如何？

实验十九　土壤阳离子交换量的测定

　　土壤是环境中污染物迁移、转化的重要场所，土壤胶体以其巨大的比表面积和带电性使土壤具有吸附性。在土壤胶体双电层的扩散层中，补偿离子可以和溶液中相同电荷的离子以离子价为依据作等价交换，称为离子交换。土壤的吸附性和离子交换性能又使它成为重金属类污染物的主要归宿。土壤阳离子交换性能是指土壤溶液中的阳离子与土壤固相的阳离子之间所进行的交换作用。它是由土壤胶体表面性质所决定的。土壤胶体指土壤中黏土矿物与腐殖酸以及相互结合形成的复杂的有机矿物质复合体，其所吸收的阳离子包括 K^+、Na^+、Mg^{2+}、NH_4^+、H^+、Al^{3+} 等。土壤交换性能对于研究污染物的环境行为有重大意义，它能调节土壤溶液的浓度，保证土壤溶液成分的多样性，从而保持土壤溶液的"生理平衡"，同时还可以保持各种养分免于被雨水淋失。土壤交换性能的分析包括阳离子交换量的测定、交换性阳离子分析及盐基饱和度的计算。

　　阳离子交换量（CEC）是指土壤胶体所能吸附的各种阳离子的总量，以每千克土壤吸附阳离子的物质的量表示（cmol/kg）。阳离子交换量的大小，可作为评价土壤保肥能力的指标。阳离子交换量是土壤缓冲性能的主要来源，是改良土壤和合理施肥的重要依据。因此，对于反映土壤负电荷总量及表征土壤性质重要指标的阳离子交换量的测定是十分重要的。土壤阳离子交换量的测定受多种因素的影响，如交换剂的性质、盐溶液浓度、pH、淋洗方法等，必须严格掌握操作技术才能获得可靠的结果。

　　联合国粮农组织规定了用于土壤分类的土壤分析的经典的中性乙酸铵法或乙酸钠法。中性乙酸铵法也是我国土壤和农化实验室所采用的常规分析方法，适于酸性和中性土壤。最近的土壤化学研究表明，对于热带和亚热带的酸性、微酸性土壤，常规方法由于浸提液 pH 值和离子强度太高，与实际情况相差较大，所得结果较实际情况偏高很多。新方法是将土壤用 $BaCl_2$ 饱和，然后用相当于土壤溶液中离子强度的 $BaCl_2$ 溶液平衡土壤，继而用 $MgSO_4$ 交换 Ba 测定酸性土壤阳离子交换量。石灰性土壤阳离子交换量的测定方法有 NH_4Cl-NH_4OAc 法、Ca（OAc）$_2$ 法和 NaOAc 法。目前应用较多且被认为较好的是 NH_4Cl-NH_4OAc 法，其测定结果准确、稳定、重现性好。NaOAc 法是目前国内广泛应用于石灰性土壤和盐碱土壤阳离子交换量测定的常规方法。随着土壤分析化学的发展，现在已有了测定土壤有效阳离子交换量的方法。如美国农业部规定用求和法测定阳离子交换量；对于以可变电荷为主的热带和亚热带地区高度风化的土壤，国际热带农业研究所建议用求和法测定土壤有效阳离子交换量（ECEC）；最近国际上又提出测定土壤有效阳离子交换量（ECEC 或 Q^+、E）和潜在阳离子交换量（PCEC 或 Q^+、P）的国际标准方法，如 ISO 11260：2018 和 ISO 13536：1995（P），这两种国际标准方法适合于各种土壤类型。

一、实验目的

（1）深刻理解土壤阳离子交换量的内涵及其环境化学意义。
（2）掌握土壤阳离子交换量的测定原理和方法。

二、实验原理

本实验采用的是快速法来测定阳离子交换量。土壤中存在的各种阳离子可被某些中性盐（$BaCl_2$）溶液中的阳离子（Ba^{2+}）等价交换（如图4-3-1）。由于在反应中存在交换平衡，交换反应实际上不能进行完全。当增大溶液中交换剂的浓度、增加交换次数时，可使交换反应趋于完全。交换离子的本性、土壤的物理状态等对交换反应的进行程度也有影响。再用强电解质（硫酸溶液）把交换到土壤中的 Ba^{2+} 交换下来，由于生成了硫酸钡沉淀，而且氢离子的交换吸附能力很强，使交换反应基本趋于完全。这样通过测定交换反应前后硫酸含量的变化，可以计算出消耗硫酸的量，进而计算出阳离子交换量。用不同方法测得的离子交换量的数值差异较大，在报告及结果应用时应注明方法。

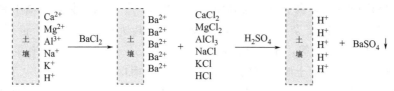

图 4-3-1　阳离子交换示意图

三、实验仪器与试剂

（一）仪器

离心机、离心管（100mL）、锥形瓶（100mL）、量筒（50mL）、移液管（10mL）、（25mL）、碱式滴定管（25mL）。

（二）试剂

（1）氯化钡溶液：称取60g氯化钡（$BaCl_2 \cdot 2H_2O$）溶于水中，转移至500mL容量瓶中，用去离子水定容。

（2）0.1%酚酞指示剂：称取0.1g酚酞溶于100mL乙醇中。

（3）硫酸溶液（0.1mol/L）：移取5.36mL浓硫酸至1000mL容量瓶中，用水稀释至刻度。

（4）标准氢氧化钠溶液（0.1mol/L）：称取2g氢氧化钠溶解于500mL煮沸后冷却的蒸馏水中，其浓度需要标定。

标定方法：各称取两份0.5000g邻苯二甲酸氢钾（预先在烘箱中105℃烘干）于250mL锥形瓶中，加100mL煮沸后冷却的蒸馏水溶解，再加4滴酚酞指示剂，用配制好的氢氧化钠标准溶液滴定至淡红色。再用煮沸后冷却的蒸馏水做一个空白实验，并从滴定邻苯二甲酸氢钾的氢氧化钠溶液的体积中扣除空白值。标准氢氧化钠溶液浓度按式（4-3-6）计算。

$$N_{NaOH} = \frac{W \times 1000}{(V_1 - V_0) \times 204.23} \tag{4-3-6}$$

式中　W——邻苯二甲酸氢钾的质量，g；

V_1——滴定邻苯二甲酸氢钾消耗的氢氧化钠溶液体积，mL；

V_0——滴定蒸馏水空白消耗的氢氧化钠体积，mL；

204.23——邻苯二甲酸氢钾的摩尔质量，g/mol。

四、实验步骤

取 4 支 100mL 离心管，分别称出其质量（准确至 0.0001g，下同）。在其中 2 支中加入 1.0g 污灌区表层风干土壤样品，其余 2 支中加入 1.0g 深层风干土壤样品，并作标记。向各管中加入 20mL 氯化钡溶液，用玻棒搅拌 4min 后，以 3000r/min 转速离心至下层土样紧实为止。弃去上清液，再加 20mL 氯化钡溶液，重复上述操作。

在各离心管内加 20mL 蒸馏水，用玻棒搅拌 1min 后，离心，弃去上清液。称出离心管连同土样的质量。移取 25.00mL 0.1mol/L 硫酸溶液至各离心管中，搅拌 10min 后，放置 20min，离心沉降，将上清液分别倒入 4 支试管中。再从各试管中分别移取 10.00mL 上清液至 4 支 100mL 锥形瓶中。同时，分别移取 10.00mL 0.1mol/L 硫酸溶液至另外 2 支锥形瓶中。在这 6 支锥形瓶中分别加入 10mL 蒸馏水、1 滴酚酞指示剂，用标准氢氧化钠溶液滴定，溶液转为红色并数分钟不褪色为终点。

五、数据处理

按式（4-3-7）计算土壤阳离子交换量（CEC）：

$$\text{CEC} = \frac{[A \times 25 - B \times (25 + G - W - W_0)] \times N}{W \times 10} \times 100 \tag{4-3-7}$$

式中　CEC——土壤阳离子交换量，cmol/kg；

　　　A——滴定 0.1mol/L 硫酸溶液消耗标准氢氧化钠溶液体积，mL；

　　　B——滴定离心沉降后的上清液消耗标准氢氧化钠溶液体积，mL；

　　　G——离心管连同土样的质量，g；

　　　W_0——空离心管的质量，g；

　　　W——称取的土样质量，g；

　　　N——标准氢氧化钠溶液的浓度，mol/L。

六、思考题

（1）解释说明两种土壤阳离子交换容量的差异。

（2）除了实验中所用的方法外，还有哪些方法可以用来测定土壤阳离子交换容量？各有什么优缺点？

（3）试述土壤的离子交换和吸附作用对污染物迁移转化的影响。

七、注意事项

（1）实验所用的玻璃器皿应洁净干燥，以免造成实验误差。

（2）离心时注意，处在对应位置上的离心管应质量接近，避免质量不平衡情况的出现。

实验二十　火焰原子吸收法测定工业固体废物中铜、锌、铅、镉的含量

固体废物是指人类在生产、消费、生活和其他活动中产生的固态、半固态废物，通俗地

说，就是垃圾。主要包括固体颗粒、垃圾、炉渣、污泥、废弃的制品、破损器皿、残次品、动物尸体、变质食品、人畜粪便等。

工业固体废物是指在工业、交通等生产活动中产生的采矿废石、选矿尾矿、燃料废渣、化工生产及冶炼废渣等固体废物，又称工业废渣或工业垃圾，包括工业废渣、废屑、污泥、尾矿等废物。

铜对人体的危害主要是有可能会导致机体铜中毒，从而影响机体的代谢功能，引发肝功能和肾功能的损伤。铜是人体必需的微量矿物质，体内铜处于正常含量范围时，对人体通常并无危害。当体内接触到大量的铜或者是体内铜代谢功能异常时，有可能会导致大量铜元素在体内的积聚。铜是一种重金属，大量的重金属积聚会导致肝脏和肾脏功能的损害。严重者会导致肝炎、低血压、昏迷、溶血、急性肾衰竭、抽搐等并发症，甚至可能发生死亡。

锌摄入量过多可致中毒，如急性锌中毒，有呕吐、腹泻等胃肠道症状；吸入工厂锌雾可能有低热及感冒样症状；慢性锌中毒可能有贫血等症状；动物实验可致肝、肾功能及免疫力受损。有些儿童玩具的涂料含锌，儿童喜欢把玩具放入口内，常因食入锌过多致中毒。

铅是广泛存在的工业污染物，能够影响人体神经系统、心血管系统、骨骼系统、生殖系统和免疫系统的功能，引起胃肠道、肝、肾和脑的疾病。

长期食用遭到镉污染的食品，可能导致痛痛病，即身体积累过量的镉损坏了肾小管功能，造成体内蛋白质从尿中流失，久而久之形成软骨征和自发性骨折。

因此，掌握火焰原子吸收法测定工业固体废物中铜、锌、铅、镉含量具有十分重要的意义。

一、实验目的

（1）掌握火焰原子吸收法测定工业固体废物中铜、锌、铅、镉含量的基本原理和操作步骤。

（2）了解铜、锌、铅、镉含量对环境评价的影响。

二、实验原理

原子吸收光谱法（AAS）是利用气态原子可以吸收一定波长的光辐射，使原子中外层的电子从基态跃迁到激发态的现象而建立的。由于各种原子中电子的能级不同，将有选择性地共振吸收一定波长的辐射光，这个共振吸收波长恰好等于该原子受激发后发射光谱的波长。当光源发射的某一特征波长的光通过原子蒸气时，即入射辐射的频率等于原子中的电子由基态跃迁到较高能态（一般情况下都是第一激发态）所需要的能量频率时，原子中的外层电子将选择性地吸收其同种元素所发射的特征谱线，使入射光减弱。特征谱线因被吸收而减弱的程度被称为吸光度 A，其在线性范围内与被测元素的含量成正比：

$$A = KC$$

式中，A 为吸光度；K 为常数（所有的常数）；C 为试样浓度。

此式就是原子吸收光谱法进行定量分析的理论基础。

将试液直接喷入火焰，在空气-乙炔火焰中，铜、锌、铅、镉的化合物解离为基态原子，并对空心阴极灯的特征辐射谱线产生选择性吸收。在给定条件下，测定铜、锌、铅、镉的吸光度。

三、实验仪器与试剂

（一）仪器

原子吸收分光光度计，附铜、锌、铅、镉空心阴极灯、乙炔钢瓶或乙炔发生器、空气压缩机（应备有除油、除水和除尘装置），仪器参数见表 4-3-2。

表 4-3-2　仪器参数

元素	铜	锌	铅	镉
测定波长/nm	324.7	213.8	283.3	228.8
通带宽度/nm	1.0	1.0	2.0	1.3
火焰性质	贫燃	贫燃	贫燃	贫燃
其他可选谱线/nm	327.4、225.8	307.6	217.0、261.4	326.2

（二）试剂

(1) 金属标准贮备液（1.000g/L）的配制：分别称取 1.000g 光谱纯金属铜、锌、铅、镉，用 20mL 优级纯硝酸（1∶1）溶解后，定容至 1000mL。

(2) 金属混合标准液的配制：用铜、锌、铅、镉的标准贮备溶液和 2% 硝酸溶液配制成含铜 20.00mg/L、锌 10.00mg/L、铅 40.00mg/L、镉 10.00mg/L 的混合标准液。

(3) 抗坏血酸（1%）：用时现配。

四、实验步骤

（一）样品采集与保存

称取 100g 试样（以干基计），置于浸取用的混合容器中，加水 1L（包括试样的含水量）。将浸取用的容器垂直固定在振荡器上，调节振荡频率为 (110±10) 次/min，振幅为 40mm，在室温下振荡 8h，静置 16h。

通过过滤装置分离固液相，并立即测定滤液的 pH 值，滤液应尽快分析，保存时间不要超过一周。

（二）空白实验

用水代样品，采用和样品相同的步骤和试剂，在测试试样的同时测定空白值。

（三）校准曲线的绘制

参考表 4-3-3，在 50mL 容量瓶中，用 0.2% HNO_3 溶液稀释混合标准溶液，配制至少 4 个工作标准溶液，其浓度范围应包括试样中铜、锌、铅、镉的浓度。

表 4-3-3　标准系列配制和浓度

混合标准液加入体积/mL	0.00	0.50	1.00	2.00	3.00	4.00
工作标准溶液中 Cd 的质量浓度/(mg/L)	0.00	0.10	0.20	0.40	0.60	1.00
工作标准溶液中 Cu 的质量浓度/(mg/L)	0.00	0.20	0.40	0.80	1.20	2.00
工作标准溶液中 Pb 的质量浓度/(mg/L)	0.00	0.40	0.80	1.60	2.40	4.00
工作标准溶液中 Zn 的质量浓度/(mg/L)	0.00	0.10	0.20	0.40	0.60	1.00

按所选的仪器工作参数调整好仪器，用硝酸溶液调零后，以低浓度到高浓度为顺序测量每份溶液的吸光度，用测得的吸光度和相对应的浓度绘制标准曲线。

（四）测定

在测量标准溶液的同时，测量空白试样。根据扣除空白后试样的吸光度，从校准曲线中查出试样中铜、铅、锌、镉的质量浓度。测定钙渣浸出液，为减少钙的干扰，须将浸出液适当稀释。测定铬渣浸出液中的铅时，除适当稀释浸出液外，为防止铅的测定结果偏低，在50mL 的试液中加入 1% 抗坏血酸 5mL，将六价铬还原成三价铬，以免生成铬酸铅沉淀。当样品中硅的浓度大于 20mg/L 时，加入钙 200mg/L，以免锌的测定结果偏低。

在测定试样的过程中，要定时复测空白和工作标准溶液，以检查基线的稳定性和仪器灵敏线是否发生了变化。

五、数据处理

浸出液中 Cu、Zn、Pb、Cd 的质量浓度 c 按式（4-3-8）计算：

$$c(\text{mg/L}) = c_1 \times \frac{V_0}{V} \tag{4-3-8}$$

式中：c_1——被测试样中金属离子的质量浓度，mg/L；

V_0——制样时定容体积，mL；

V——试样的体积，mL。

六、思考题

（1）试分析讨论固体废物中铜、锌、铅、镉的主要来源。

（2）哪些因素会干扰固体废物中铜、锌、铅、镉的测定，如何消除？

第四节　化学物质的生态效应实验

实验二十一　底泥对苯酚的吸附作用

底泥/悬浮颗粒物是水中污染物的源和汇。水体中有机污染物的迁移转化途径很多，如挥发、扩散、化学或生物降解等，其中底泥/悬浮颗粒物的吸附作用对有机污染物的迁移、转化、归趋及生物效应有重要影响，在某种程度上起着决定作用。底泥对有机物的吸附主要包括分配作用和表面吸附。

苯酚是化学工业的基本原料，也是水体中常见的有机污染物。底泥对苯酚的吸附作用与其组成、结构等有关。吸附作用的强弱可用吸附系数表示。探讨底泥对苯酚的吸附作用对了解苯酚在水/沉积物多介质的环境化学行为，乃至水污染防治都具有重要的意义。

本实验以底泥为吸附剂，吸附水中的苯酚，绘制吸附等温线后，用回归法求出底泥对苯酚的吸附常数。

一、实验目的

（1）绘制底泥对苯酚的吸附等温线，求出吸附常数。

(2) 了解水体中底泥的环境化学意义及其在水体自净中的作用。

二、实验原理

实验研究底泥对一系列浓度苯酚的吸附情况，计算平衡浓度和相应的吸附量，通过绘制等温吸附曲线，分析底泥的吸附性能和机理。

本实验采用 4-氨基安替比林法测定苯酚。即在 pH 10.0 ± 0.2 介质中，在铁氰化钾存在下，苯酚与 4-氨基安替比林反应，生成橙色的吲哚酚安替比林染料，其水溶液在波长 510nm 处有最大吸收。用 2cm 比色皿测量时，苯酚的最低检出浓度为 0.1mg/L。

三、实验仪器与试剂

（一）仪器

恒温调速振荡器、低速离心机、可见光分光光度计、碘量瓶（100mL）、离心管（50mL）、比色管（50mL）、移液管（0.5mL、1.0mL、2mL、5mL、10mL、20mL）。

（二）试剂

（1）底泥样品制备及表征：采集河道的表层底泥，去除沙砾和植物残体等大块物，于室温下风干；用瓷研钵捣碎，过 100 目筛（＜0.15mm），充分摇匀，装瓶备用。用固体总有机碳分析仪测定土壤中有机碳含量（f_{oc}）。

（2）无酚水：于 1L 水中加入 0.2g 经 200℃活化 0.5h 的活性炭粉末，充分振荡后，放置过夜。用双层中速滤纸过滤，或加氢氧化钠使水呈碱性，并滴加高锰酸钾溶液至紫红色，移入蒸馏瓶中加热蒸馏，收集流出液备用。本实验应使用无酚水。

注：无酚水应贮备于玻璃瓶中，取用时应避免与橡胶制品（橡皮塞或乳胶管）接触。

（3）淀粉溶液：称取 1g 可溶性淀粉，用少量水调成糊状，加沸水至 100mL，冷却，置于冰箱保存。

（4）溴酸钾-溴化钾标准参考溶液（$c_{1/6KBrO_3}$＝0.1mol/L）：称取 2.784g 溴酸钾溶于水中，加入 10g 溴化钾，使其溶解，移入 1000mL 容量瓶中，稀释至标线。

（5）碘酸钾标准参考溶液（$c_{1/6KIO_3}$＝0.0125mol/L）：称取预先在 180℃烘干的碘酸钾 0.4458g 溶于水中，移入 1000mL 容量瓶中，稀释至标线。

（6）硫代硫酸钠标准溶液（$c_{Na_2S_2O_3}$≈0.0125mol/L）：称取 3.1g 硫代硫酸钠溶于煮沸后冷却的水中，加入 0.2g 碳酸钠，稀释至 1000mL，临用前，用碘酸钾标准参考溶液标定。

标定方法：取 10.0mL 碘酸钾溶液置于 250mL 碘量瓶中，加水稀释至 100mL，加 1g 碘化钾，再加 5mL（1+5）硫酸，加塞，轻轻摇匀。置于暗处放置 5min，用硫代硫酸钠溶液滴定至淡黄色，加 1mL 淀粉溶液，继续滴定至蓝色刚褪去为止，记录硫代硫酸钠溶液用量。按式（4-4-1）计算硫代硫酸钠溶液物质的量浓度（mol/L）：

$$c_{Na_2S_2O_3}=\frac{0.0125V_4}{V_3} \tag{4-4-1}$$

式中　$c_{Na_2S_2O_3}$——硫代硫酸钠物质的量浓度，mol/L；

　　　V_3——硫代硫酸钠溶液消耗量，mL；

　　　V_4——移取碘酸钾标准参考溶液量，mL；

　　　0.0125——碘酸钾标准参考溶液浓度，mol/L。

(7) 苯酚标准储备液：称取 1.00g 无色苯酚溶于水中，移入 1000mL 容量瓶中，稀释至标线。在冰箱内保存，至少稳定 1 个月。

标定方法：吸取 10.00mL 苯酚储备液于 250mL 碘量瓶中，加水稀释至 100mL，加 10.0mL 0.1mol/L 溴酸钾-溴化钾溶液，立即加入 5mL 盐酸，盖好瓶塞，轻轻摇匀，在暗处放置 10min。加入 1g 碘化钾，盖好瓶塞，再轻轻摇匀，在暗处放置 5min。用 0.0125mol/L 硫代硫酸钠标准溶液滴定至淡黄色，加入 1mL 淀粉溶液，继续滴定至蓝色刚好褪去，记录用量。同时以水代替苯酚储备液做空白实验，记录硫代硫酸钠标准溶液滴定用量。苯酚储备液的质量浓度由式（4-4-2）计算：

$$\rho_{苯酚}=\frac{(V_1-V_2)c\times 15.68}{V} \tag{4-4-2}$$

式中 $\rho_{苯酚}$——苯酚储备液的质量浓度，mg/mL；

V_1——空白实验中硫代硫酸钠标准溶液滴定用量，mL；

V_2——滴定苯酚储备液时，硫代硫酸钠标准溶液滴定用量，mL；

V——取用苯酚储备液体积，mL；

c——硫代硫酸钠标准溶液浓度，mol/L；

15.68——1/6 苯酚摩尔质量，g/mol。

(8) 苯酚标准中间液（使用时当天配制）：取适量苯酚贮备液，用水稀释，配制成 10μg/mL 苯酚中间液。

(9) 苯酚吸附使用液（2000μg/mL）：称取 2.00g 无色苯酚溶于水中，移入 1000mL 的容量瓶中，稀释至标线。

(10) 缓冲溶液（pH 约为 10）：称取 20g 氯化铵溶于 100mL 氨水中，加塞，置于冰箱中保存。

(11) 2% 4-氨基安替比林溶液：称取 4-氨基安替比林（$C_{11}H_{13}N_3O$）2g 溶于水，稀释至 100mL，置于冰箱中保存。可使用 1 周。

(12) 8% 铁氰化钾溶液：称取 8g 铁氰化钾 $\{K_3[Fe(CN)_6]\}$ 溶于水，稀释至 100mL。置于冰箱内可保存 1 周。

四、实验步骤

（一）标准曲线的绘制

在 9 支 50mL 比色管中分别加入 0.00、1.00、3.00、5.00、7.00、10.00、12.00、15.00、18.00mL 浓度为 10μg/mL 的苯酚标准液，用水稀释至刻度。加 0.5mL 缓冲溶液，混匀，此时 pH 为 10.0±0.2。加 4-氨基安替比林溶液 1.0mL，混匀，再加 1.0mL 铁氰化钾溶液，充分混匀后，放置 10min，立即在 510nm 波长处，以蒸馏水作为参比，用 2cm 比色皿，测量吸光度，记录数据，经空白校正后，绘制吸光度对苯酚质量浓度（μg/mL）的标准曲线。

（二）吸附实验

取 6 支干净的 100mL 碘量瓶，分别在每个瓶内放入 1.0g（精确到 0.0001g，以下同）左右的沉积物样品。然后按表 4-4-1 所给参数加入浓度为 2000μg/mL 的苯酚使用液和无酚水，加塞密封并摇匀后，将瓶子放入振荡器中，在（25±1.0）℃下，以 150~175r/min 的转速振荡 8h，静置 30min 后，在低速离心机上以 3000r/min 速度离心 5min，移出上清液

10mL 至 50mL 容量瓶中,用蒸馏水定容至刻度,摇匀,然后移出数毫升(视平衡浓度而定)至 50mL 比色管中,用水稀释至刻度。按与绘制标准曲线相同的步骤测定吸光度,从标准曲线上查出苯酚的质量浓度,将数据记录入表 4-4-1。

表 4-4-1 苯酚加入浓度系列

项目	1	2	3	4	5	6
苯酚吸附使用液体积/mL	1.0	3.0	6.0	12.5	20.0	25.0
无酚水体积/mL	24	22	19	12.5	5	0
起始浓度 ρ_0/(mg·L^{-1})	80	240	480	1000	1600	2000
取上清液的体积/mL	2.00	1.00	1.00	1.00	0.50	0.50
稀释倍数	125	250	250	250	500	500
吸光度						
平衡浓度 ρ_e/(mg·L^{-1})						
吸附量 Q/(mg·kg^{-1})						

五、数据处理

(1) 计算平衡浓度(ρ_e)及吸附量(Q)。由式(4-4-3)和式(4-4-4)计算。

$$\rho_e = \rho_1 n \tag{4-4-3}$$

$$Q = \frac{(\rho_0 - \rho_e)V}{m} \tag{4-4-4}$$

式中　ρ_0——起始浓度,μg/mL;

ρ_e——平衡浓度,μg/mL;

ρ_1——在标准曲线上查得的测量浓度,mg/L;

n——溶液的稀释倍数;

V——吸附实验中所加苯酚溶液的体积,mL;

m——吸附实验所加底泥样品的质量,g;

Q——苯酚在底泥样品上的吸附量,mg/kg。

(2) 利用平衡浓度和吸附量数据绘制苯酚在底泥上的吸附等温曲线。

(3) 利用 Freundlich 吸附方程 $Q = K\rho^{1/n}$,通过回归分析求出方程中的常数 K 及 n。

六、思考题

(1) 影响底泥对苯酚吸附系数大小的因素有哪些?

(2) 哪种吸附方程更能准确描述底泥对苯酚的等温吸附曲线?

实验二十二　水中有机物的挥发速率

水环境中有机污染物随自身的物理化学性质和环境条件的不同而进行不同的迁移转化过程,诸如挥发、微生物降解、光解以及吸附等。近年来研究表明,自水体挥发进入空气是疏水性有机污染物特别是高挥发性有机污染物的主要迁移途径。

水中有机污染物的挥发符合一级动力学方程，其挥发速率常数可通过实验求得，其数值的大小受温度、水体流速、风速和水体组成等因素的影响。测定水中有机物的挥发速率，对研究其在环境中的归宿具有重要的意义。

一、实验目的

掌握测定水中溶解的有机物质挥发速率的实验方法。

二、实验原理

水中溶解的有机物的挥发符合一级动力学方程，即式（4-4-5）：

$$\frac{dc}{dt} = -K_v c \tag{4-4-5}$$

式中　K_v——挥发速率常数，s^{-1}；

　　　c——水中有机物的浓度，$g \cdot L^{-1}$；

　　　t——挥发时间，s。

对式（4-4-5）积分可得式（4-4-6）：

$$\ln \frac{c_0}{c} = K_v t \tag{4-4-6}$$

由此可求得有机物质挥发掉一半所需的时间（$t_{1/2}$）为式（4-4-7）：

$$t_{1/2} = \frac{0.693}{K_v} \tag{4-4-7}$$

若 L 为溶液在一定截面积的容器中的高度，则传质系数 K 与挥发速率常数 K_v 的关系如式（4-4-8）：

$$K_v = \frac{K}{L} \tag{4-4-8}$$

因此，只要求得某种化合物的挥发速率常数 K_v，就能求得传质系数 K。

三、实验仪器与试剂

（一）仪器

紫外分光光度计、电子天平、称量瓶、烧杯、容量瓶、尺子。

（二）试剂

甲苯、甲醇（均为分析纯）。

四、实验步骤

（1）储备液的配制：准确称取甲苯 2.5000g，置于 250mL 的容量瓶中，用甲醇稀释到刻度，溶液质量浓度约为 10mg/mL。

（2）中间液的配制：取上述储备液 5mL 置于 250mL 的容量瓶中，用水稀释至刻度，溶液质量浓度为 200mg/L。

（3）标准曲线的绘制：取甲苯中间液 0.25、0.5、1.0、1.5 和 2.0mL 于 10mL 的容量瓶内，用水稀释至刻度，其质量浓度分别为 5、10、20、30 和 40mg/L。将该组溶液用紫外分光光度计于波长 205nm 处测定吸光度，以吸光度对质量浓度作图，可得到甲苯的标准曲线。

（4）将剩余的甲苯的中间液分别倒入 2 个烧杯内，量出溶液高度 L，并记录时间。让其自然挥发，每隔 10min 取样一次，每次取 1.0mL，用去离子水定容至 10mL，测定吸光度，测定波长为 205nm，共测 10 个点。

五、数据处理

（一）求半衰期 $t_{1/2}$ 和甲苯的挥发速率常数

从标准曲线上查得甲苯不同反应时间在溶液中的质量浓度，绘制 $\ln(c_0/c)$-t 关系曲线，从其斜率（K_v）即可求得半衰期：

$$t_{1/2} = \frac{0.693}{K_v}$$

（二）求传质系数 K

由式（4-4-8）$K_v = \dfrac{K}{L}$ 即可求出化合物的传质系数 K。

六、思考题

影响环境中有机污染物挥发的因素有哪些？

实验二十三　底泥中铬的简单状态鉴别

一、实验目的

（1）学习底泥中不同状态铬的提取技术。
（2）测定不同状态铬的含量。

二、实验原理

水体底泥中铬污染多数是由于电镀、制革和印染等污水排放所引起的。在还原性及中性水体中，污染源排出的六价铬很容易被还原为三价铬。在水体中三价铬主要以氢氧化物形式存在。它很容易吸附在悬浮物粒子上随水迁移，在迁移过程中逐步沉积进入底泥。

底泥中的铬主要有被无机交换剂和有机交换剂交换吸附的铬，以氢氧化物形式沉积的铬及可被强酸置换的腐殖质吸附的铬，还有与高分子量腐殖质形成稳定配合物的铬，以及形成最稳定结构的复合铬。

上述几种主要状态中，可交换态铬可被一价或二价阳离子，如 NH_4^+、Na^+、Ba^{2+}，以及某些混合电解质溶液交换出来，以氢氧化物形式存在的铬和吸附态铬可用稀酸溶解出来，与高分子量腐殖质形成稳定配合物的铬可用稀碱提取，最稳定的残渣部分可用硝酸和高氯酸消解溶出。

本实验以受铬污染的底泥为试样，依次用海水、稀酸、稀碱、硝酸和高氯酸对试样中的铬进行逐步溶出，然后测出各溶出液的含铬量。

三、实验仪器与试剂

（一）仪器

721（或72）型分光光度计、控温电炉、电动离心机、水浴锅、比色管（25mL）、离心管（50mL）、锥形瓶（100mL）、容量瓶（100mL）、量筒（25mL和50mL）、移液管（2mL、5mL和10mL）。

（二）试剂

饱和 NaOH 溶液、0.1mol/L NaOH 溶液、0.1mol/L HNO_3 溶液、0.5mol/L H_2SO_4 溶液（体积比为1:1)、4% $KMnO_4$ 溶液、95%乙醇、天然海水（或模拟海水，含 NaCl 3.3%）、分析纯浓 HNO_3、分析纯浓 $HClO_4$、甲基橙指示剂（质量浓度为0.1%）、Cr(Ⅵ)贮备液（160μg/mL）、Cr(Ⅵ)标准液（4μg/mL）。

显色剂：称0.2g 二苯碳酰二肼溶于50mL 丙酮中，加水稀释至100mL，摇匀，贮于棕色瓶置于冰箱中保存，色变深后不能使用。

底泥：风干后过150目筛。

四、实验步骤

称0.5g（准确到0.0001g）左右底泥两份，分别放入两个质量接近的离心管中。

（一）交换态铬

往管内各加入20mL 天然海水，搅拌半小时。把离心管放在离心机对称位置上离心10min，将上部清液倒入100mL 锥形瓶内，再向离心管内加入20mL 蒸馏水，搅拌均匀后离心10min，上部清液合并入100mL 锥形瓶内。离心管内泥样供下述实验。

（二）酸溶态铬

往离心管内加入20mL 0.1mol/L 硝酸溶液，搅拌半小时。离心10min，将上部清液倒入100mL 锥形瓶内，再用20mL 蒸馏水洗涤泥样，离心分离出清液，合并到酸溶出液中。离心管内泥样供下述实验。

（三）碱溶态铬

往离心管内加入20mL 0.1mol/L 氢氧化钠溶液，把离心管放在80℃水浴锅内搅拌半小时。离心10min，上层清液倒入100mL 锥形瓶内，再用20mL 蒸馏水洗涤泥样，离心分离出洗涤液再并入碱溶出液中。泥样继续供下述实验。

（四）残渣态铬

用40mL 蒸馏水把离心管内泥样定量地洗入100mL 锥形瓶中。

（五）消化处理

往各锥形瓶中加入1mL 浓硝酸，再把全部锥形瓶放在控温电炉上将试液浓缩到10mL 左右。各加入4mL 高氯酸，瓶口盖上表面皿，继续消化到水样清亮、泥样变白、瓶内充满白烟为止。残渣如不变白，可补加4mL 浓硝酸继续消化至残渣变白。取下锥形瓶，冷却后消化液用蒸馏水定量转入100mL 容量瓶内，并定容到100mL，再分别从其中移出5mL 溶液到干净的100mL 锥形瓶内，并加15mL 蒸馏水，各瓶做好标记并加入玻璃珠。然后取出1瓶溶液，加1滴甲基橙指示剂，逐滴加入饱和氢氧化钠溶液至溶液刚好变黄为止，再慢慢滴加 0.5mol/L 硫酸到刚好变红，如此逐个将 8 个瓶中溶液的 pH 值调好，再分别加入一滴 1∶1 硫酸溶液。

（六）氧化处理

用蒸馏水将上述 8 瓶溶液的体积调整到 25mL 左右，再放在电炉上加热，沸腾后各加入 2 滴 4% $KMnO_4$ 溶液。继续煮沸 2min，趁热沿瓶壁加入 1mL 乙醇，再煮沸 2min。取下锥形瓶，冷却后将各瓶溶液的 pH 值调到中性（pH 试纸指示），然后过滤到 25mL 比色管中，用蒸馏水洗涤锥形瓶和漏斗，使管内滤液恰为 25mL。

（七）标准溶液制备

取 6 支 25mL 比色管，分别加入 0、0.5、1.0、1.5、2.0、2.5mL Cr(Ⅵ) 标准液，再用蒸馏水稀释到刻度。

（八）测定

每 6 支比色管为一批，分别加入 1.25mL 显色剂，摇匀。显色 10min。用 2cm 比色皿盛显色液，在分光光度计上于 540nm 波长处以 0 号标准溶液为参比测定各显色液的吸光度。测完把玻璃仪器洗净，再泡入硝酸洗液。

五、数据处理

(1) 以标准溶液中 Cr(Ⅵ) 质量浓度为横坐标，吸光度为纵坐标，作标准曲线。

(2) 根据各状态铬显色液的吸光度查标准曲线得出 Cr(Ⅵ) 的质量浓度，再进一步算出每千克泥样含铬的质量，最后求出它们的平均值。将数据填入下列表 4-4-2 和表 4-4-3 中。

表 4-4-2　标准曲线数据

瓶号	1	2	3	4	5	6
Cr(Ⅵ)加入量/μg						
吸光度						

表 4-4-3　状态和含量数据

项目	交换态		酸溶态		碱溶态		残渣态	
	1	2	1	2	1	2	1	2
吸光度								
Cr(Ⅵ)质量/μg								
含铬质量浓度/(mg·L^{-1})								
平均含铬质量浓度/(mg·L^{-1})								

注：消化和氧化过程必须在通风橱内进行，并注意勿使强酸和强碱溶液溅出。

六、思考题

(1) 底泥中的铬可能以哪几种状态存在？说明原因。

(2) 由实验结果说明所用底泥中铬的主要存在形态。

实验二十四　沉积物中重金属的存在形式和迁移规律的研究

一、实验目的

(1) 了解沉积物中重金属的存在形式及其迁移规律。
(2) 掌握沉积物中重金属不同存在形态的化学提取操作技术。
(3) 掌握原子吸收法的基本原理以及正确操作技能。

二、实验原理

（一）存在的形态类型

分析污染物在水体中的迁移转化规律，首先就要了解污染物在水体中以何种形式存在以及各存在形态之间的关系，对重金属污染物的研究也不例外。汤鸿霄提出"所谓形态，实际上包括价态、化合态、结合态和结构态 4 个方面，有可能表现出来不同的生物毒性和环境行为"，这里所分析的存在形态主要指重金属在水体中的结合态。水体中重金属存在形态可分为溶解态和颗粒态，即用 $0.45\mu m$ 滤膜过滤水样，滤水中的为溶解态（溶解于水中），原水样中未过滤的为颗粒态（包括存在于悬移质中的悬移态及存在于表层沉积物中的沉积态）。用 Tessier 等提出的逐级化学提取法又可将颗粒态重金属继续划分为以下 5 种存在形态：一是可交换态，指吸附在悬浮沉积物中的黏土、矿物、有机质或铁锰氢氧化物等表面上的重金属；二是碳酸盐结合态，指结合在碳酸盐沉淀上的重金属；三是铁锰水合氧化物结合态，指水体中重金属与水合氧化铁、氧化锰生成和结合的部分；四是有机硫化物和硫化物结合态，指颗粒物中的重金属以不同形式进入或包裹在有机颗粒上，同有机质发生螯合反应或生成硫化物；五是残渣态，指重金属存在于石英、黏土、矿物等结晶矿物晶格中的部分。

（二）迁移性质

不同存在形态的重金属在水体中的迁移性质不同。溶解态重金属对人类和水生生态系统的影响最直接，是人们判断水体中重金属污染程度的常用依据之一。颗粒态重金属组成复杂，其形态性质各不相同。可交换态重金属是最不稳定的，只要环境条件变化，极易溶解于水或被其他极性较强的离子交换，是影响水质的重要组成部分。碳酸盐结合态重金属在环境变化，特别是 pH 值变化时最易重新释放进入水体。铁锰水合氧化物结合态重金属在环境变化时也会部分释放。有机硫化物和硫化物结合态重金属不易被生物吸收，较稳定。残渣态重金属最稳定，在相当长的时间内不会释放到水体中。

（三）迁移规律研究方法

不同存在形态的重金属，从所结合的载体上分离下来的化学条件和难易程度也不同，即稳定性存在差异，因此其对水体造成的污染程度也不相同。不同的重金属污染物在水体中存在形态的分布规律存在差异，可以通过研究它们之间的分布差异以及相互转化过程，研究重金属迁移转化过程，并作为判断其对水体危害的依据。分析沉积物重金属污染问题时，仅认识到重金属的总量是不够的，还需要分析其中的各组分含量和分布规律，进而讨论沉积物中重金属污染物的污染性质、转化机理以及对水体的潜在污染等问题。在研究整个水体中重金属污染问题时，也常使用该方法分析重金属水相和固相相互迁移的主要形式，据此得出不同形态重金属在水体中迁移的动态转换规律以及最终归宿等。

三、实验仪器与试剂

(一) 仪器

3200 原子吸收分光光度计（附铜、铅、镉、锌空心阴极灯）、电动离心机及塑料离心管、调温磁力搅拌器及磁搅拌子、石英烧杯（50mL）、容量瓶（50mL、25mL）、移液管（25mL、10mL、5mL、2mL）、聚四氟乙烯烧杯（50mL）、塑料量筒（25mL、10mL）、量筒（50mL、10mL）、电热板、蚌式采泥器、磁研钵、玛瑙研钵、尼龙筛（100目）、重蒸馏水。

(二) 试剂

1mol/L $MgCl_2$（用 HCl 调 pH=7）、1mol/L NaAc（用 HAc 调 pH=5）、0.02mol/L HNO_3、浓 HF (GR)、浓 $HClO_4$ (GR)、0.04mol/L $NH_2OH·HCl$（溶于 25% 的 HAc 中）、30% H_2O_2（用 HNO_3 调 pH=2）、3.2mol/L NH_4Ac（溶于 20% HNO_3 中）、金属（Cu、Pb、Cd、Zn）标准贮备溶液（1000mg/L）。

混合标准溶液：用 0.2% 硝酸稀释金属标准贮备溶液配制而成，使配成的混合标准溶液每毫升含 Cu、Pb、Cd、Zn 分别为 50.0、100.0、10.0 和 10.0μg。

四、实验步骤

(一) 采样及样品预处理

用蚌式采泥器（也可直接用其他容器采样）抓取表层沉积物，出水后迅速用不锈钢刀刮去表层，取其中心部分数百克样品，封存在塑料袋中，并储存在冰箱中。

使用前从塑料袋中取出泥样，放在玻璃表面皿或培养皿中，在无污染的空气流通处晾干，阴干的样品先在磁研钵中研细，再用玛瑙研钵研细到 80~100 目，贮于干燥器中备用。

(二) 沉积物中不同存在形态重金属的化学连续提取法

1. 可交换态（吸附态）

准确称取 1.000g 样品于石英烧杯中，加入 25mL 1mol/L $MgCl_2$ 搅拌一小时，离心沉降。离心液转移到 50mL 容量瓶中，残渣用 10mL 蒸馏水洗涤，离心液合并到 50mL 容量瓶中，稀释到刻度，供测定可交换重金属含量用。保留残渣 1 供提取碳酸盐结合态重金属用。

2. 碳酸盐结合态

残渣 1 分次用 20mL 1mol/L NaAc（预先用 HAc 调 pH=5.00）转移到石英烧杯中，搅拌提取 5h，离心沉降。离心液转移到 50mL 容量瓶中，残渣用 10mL 重蒸馏水洗涤，离心沉淀，离心液合并到 50mL 容量瓶中，用重蒸馏水稀释到刻度，供测定碳酸盐结合态重金属含量用。保留残渣 2，供提取铁锰水合氧化物结合态重金属用。

3. 铁锰水合氧化物结合态

残渣 2 用 40mL 0.04mol/L $NH_2OH·HCl$（溶于体积分数为 25% 的 HAc 中）分次转移到石英烧杯中，在（96±3）℃水浴中搅拌提取 5h，离心沉降。离心液转移到 50mL 容量瓶中，残渣用 10mL 重蒸馏水洗涤，离心沉降，离心液并到 50mL 容量瓶中，用水稀释到刻度，供测定铁锰水合氧化物结合态重金属含量用。保留残渣 3，供提取有机硫化物和硫化物结合态重金属用。

4. 有机硫化物和硫化物结合态

残渣 3 用 6mL 0.02mol/L HNO_3 转移到石英烧杯中，再加入 10mL 30% H_2O_2（预先

用 HNO_3 调 pH=2），在（85±2）℃水溶中搅拌提取 2h，继续搅拌 3h，取出冷却，再加入 10mL 3.2mol/L NH_4AC（溶于体积分数为 20% 的 HNO_3 中）和 15mL 重蒸馏水，搅拌 1h，离心沉降。离心液转移到 50mL 容量瓶中，残渣用 10mL 重蒸馏水洗涤，离心沉降，离心液合并到 50mL 容量瓶中，用重蒸馏水稀释到刻度，溶液供测定有机硫化物和硫化物结合态重金属含量用。保留残渣 4 供消解用。

5. 残渣态

残渣 4 用 15mL $HClO_4$ 分次转移到聚四氟乙烯（50mL）烧杯中，在通风橱中加入 5mL HF（用塑料量筒量取），用电热板（或电炉垫以石棉网）加热消化（注意请勿蒸干）。反复用上述比例 $HClO_4$ 和 HF 处理，直到沉积物消化完全，蒸发近干。加入 1mL $HClO_4$ 及少量重蒸馏水（10~20mL），微热，这时溶液应该透明、澄清。液体转移到 50mL 容量瓶中，用少许微酸性（$HClO_4$ 酸化）蒸馏水洗涤聚四氟乙烯烧杯数次，并转入容量瓶中，继续稀释到刻度。所得溶液供测定残渣态重金属含量用。

（三）测定

1. 标准溶液的配制

分别吸取混合标准溶液 0、0.25、0.50、1.50、2.50 和 5.00mL 放入 6 个 50mL 容量瓶中，用 0.2% HNO_3 稀释定容。此混合标准系列溶液各金属质量浓度见表 4-4-4。

表 4-4-4 混合标准系列溶液的配制和浓度

混合标准溶液体积/mL		0	0.25	0.50	1.50	2.50	5.00
混合标准系列溶液各金属质量浓度/(mg/L)	Cu	0	0.25	0.50	1.50	2.50	5.00
	Pb	0	0.50	1.00	3.00	5.00	10.0
	Cd	0	0.05	0.10	0.30	0.50	1.00
	Zn	0	0.05	0.10	0.30	0.50	1.00

注：定容体积为 50mL。

2. 标样及样品溶液的测定

将标准系列溶液与待测样品溶液在相同条件下分别进行原子吸收测定，原子吸收的测定条件如表 4-4-5 所示，将所得的吸光度值对质量浓度作图，得到工作曲线。从该曲线上求出待测元素的质量浓度。

表 4-4-5 原子吸收测定条件

元素	Cu	Pb	Cd	Zn
波长/nm	324.7	283.3	228.8	213.8
灯电流/mA	10	10	8	10
狭缝位置	2	2	2	2
阻尼位置	1	1	1	1
增益位置	2.9	4.0	4.5	4.5
燃气	乙炔	乙炔	乙炔	乙炔
助燃气	空气	空气	空气	空气
火焰类型	氧化型	氧化型	氧化型	氧化型

五、数据处理

$$沉积物中铜、铅、镉、锌的质量分数(\mathrm{mg/kg}) = \frac{MV}{W} \quad (4\text{-}4\text{-}9)$$

式中　M——标准曲线上查出的样品质量浓度，mg/L；
　　　V——铜、铅、镉、锌各形态的最后定容的体积，mL；
　　　W——沉积物的质量，g。

六、思考题

(1) 为什么要对沉积物进行形态分析？
(2) 为什么要重视提取剂的 pH？
(3) 本实验有哪些地方需特别注意？
(4) 根据实验测定结果，评价沉积物的铜、铅、镉、锌污染状况及其行为。
(5) 进行原子吸收光谱分析，一般要做哪些条件实验？

七、注意事项

(1) 由于沉积物来源不同，铜、铅、镉、锌含量不同，因而标准曲线的绘制可视具体情况作适当调整。

(2) 用 HF 分解残渣时，加热不应超过 250℃，以免聚四氟乙烯分解产生有毒的含氟异丁烯气体。最好在铂器皿中进行试样分解。另外，HF 对人体有毒和有腐蚀性，使用时应注意勿吸入 HF 蒸气，也不可接触 HF，HF 接触皮肤后会引起灼伤溃烂，且不易愈合。

(3) 测定试样时，应对不同形态作不同的空白对照试验。

实验二十五　氢化原子荧光光度法测定食品中总砷的含量

砷是地壳的一个天然成分，并广泛分布在空气、水和陆地等环境中。砷的蒸气具有一股难闻的大蒜臭味。金属砷很容易与氟、氧发生反应，在加热条件下能与大多数金属、非金属发生反应。砷不溶于水，但溶于硝酸、王水和强碱。

三氧化二砷（砒霜）的毒性很强，进入人体后能破坏某些细胞呼吸酶，使组织细胞不能获得氧气而死亡；还能强烈刺激胃肠黏膜，使黏膜溃烂、出血；亦可破坏血管，使人体出血，破坏肝脏，严重的话人会因呼吸和循环衰竭而死。

对公共卫生最大的砷威胁来自受污染的地下水。在工业上，砷用作合金添加剂，它同样用于玻璃、涂料、纺织品、纸张、金属黏合剂、木材防腐剂和弹药的处理。砷还用于制革工艺，并在有限程度上用于杀虫剂、饲料添加剂和药物。吸烟也可接触烟草中含有的天然无机砷，因为烟草植物主要是从土壤中摄取天然存在的砷。急性砷中毒的早期症状包括呕吐、腹部疼痛和腹泻，随后是四肢麻痹和刺痛，肌肉痉挛，在极端情况下会发生死亡。

长期接触饮用水和食品中含有的砷可导致癌症和皮肤病变。砷是世卫组织列出的引起重大公共卫生关注的 10 种化学品之一。目前饮用水中建议的砷含量限值为 $10\mu\mathrm{g/L}$。因此，掌握氢化原子荧光光度法测定食品中总砷含量具有十分重要的意义。

一、实验目的

（1）掌握氢化原子荧光光度法测定食品中总砷含量的基本原理和操作步骤。

（2）了解砷对人类健康的影响。

二、实验原理

食品试样经湿消解或干灰化后，加入硫脲使五价砷预还原为三价砷，再加入硼氢化钠使其还原生成砷化氢，砷化氢由氩气载入石英原子化器中被分解为原子态砷，在特制砷空心阴极灯的发射光激发下产生原子荧光，其荧光强度在固定条件下与被测液中砷浓度成正比，与标准系列比较定量。

三、实验仪器与试剂

（一）仪器

原子荧光光度计。

（二）试剂

（1）氢氧化钠溶液（2g/L、100g/L）、硫脲溶液（50g/L）、硫酸溶液（1+9）。

（2）硼氢化钠溶液（10g/L）：称取硼氢化钠10.0g，溶于1000mL 2g/L氢氧化钠溶液中，混匀。

（3）砷标准储备液（含砷0.1g/L）：精确称取于100℃干燥2h以上的三氧化二砷0.1320g，加100g/L氢氧化钠溶液溶解，用适量水转入1000mL容量瓶中，加（1+9）硫酸25mL，用水定容至刻度。

（4）砷标准使用液（含砷1μg/mL）：吸取1.00mL砷标准储备液于100mL容量瓶中，用水稀释至刻度。此液应当日配制使用。

（5）湿消解试剂：硝酸、硫酸、高氯酸。

（6）干灰化试剂：六水硝酸镁（150g/L）、氯化镁、盐酸（1+1）。

（7）砷标准系列溶液：取25mL容量瓶或比色管6支，依次准确加入1μg/mL砷标准使用液0、0.05、0.2、0.5、2.0、5.0mL，各加（1+9）硫酸12.5mL，50g/L硫脲2.5mL，补加水至刻度，混匀备测。相当于砷质量浓度分别为0.0、2.0、8.0、20.0、80.0、200.0ng/mL。

四、实验步骤

（一）试样消解

（1）湿消解：称固体试样1～2.5g，称液体试样5～10g（或mL）（精确至小数点后第二位），置入50～100mL锥形瓶中，同时做两份试剂空白。加硝酸20～40mL，硫酸1.25mL，摇匀后放置过夜，置于电热板上加热消解。若消解液处理至10mL左右时仍有未分解物质或色泽变深，取下放冷，补加硝酸5～10mL，再消解至10mL左右观察，如此反复两三次，注意避免炭化。如仍不能消解完全，则加入高氯酸1～2mL，继续加热至消解完全后，再持续蒸发至高氯酸的白烟散尽，硫酸的白烟开始冒出。冷却，加水25mL，再蒸发至冒硫酸白烟。冷却，用水将内容物转入25mL容量瓶或比色管中，加入50g/L硫脲2.5mL，补水至刻度并混匀，备测。

（2）干灰化：称取1~2.5g固体试样（精确至小数点后第二位）于50~100mL坩埚中，同时做两份试剂空白。加150g/L硝酸镁10mL混匀，低热蒸干，将1g氧化镁仔细覆盖于渣上，于电炉上炭化至无黑烟，移入550℃高温灰化炉灰化4h，取出冷却，小心加入（1+1）盐酸10mL，以中和氧化镁并溶解灰分，转入25mL容量瓶或比色管中，向容量瓶或比色管中加入50g/L硫脲2.5mL，另用（1+9）硫酸分次刷洗坩埚后转出合并，直至25mL刻度，混匀备测。

（二）测定

（1）仪器参考条件：光电倍增管电压为400V；砷空心阴极灯电流为35mA；原子化器温度为820~850℃，高度为7mm；氩气流速为600mL/min；测量方式为荧光强度或浓度直读；读数方式为峰面积；读数延迟时间为1s；读数时间为15s；硼氢化钠溶液加入时间为5s；标液或样液加入体积为2mL。

（2）浓度测量方式：如直接测荧光强度，则在开机并设定好仪器条件后，预热稳定约20min，按"B"键进入空白值测量状态，连续用标准系列的"0"管进样，待读数稳定后，按空挡键记录下空白值（即让仪器自动扣底）即可开始测量。先依次测标准系列（可不再测"0"管），标准系列测完后应仔细清洗进样器，记录（或打印）下测量数据。

（3）仪器自动方式：利用仪器提供的软件功能进行浓度直读测定，在开机、设定条件和预热后，还需输入必要的参数，即试样量（g或mL）、稀释体积（mL）、进样体积（mL）、结果的浓度单位、标准系列各点的重复测量次数、标准系列的点数（不计零点）及各点的浓度值。首先进入空白值测量状态，连续用标准系列的"0"管进样以获得稳定的空白值并执行自动扣底后，再依次测标准系列（此时"0"管需再测一次）。在测样液前，须再进入空白值测量状态，先用标准系列"0"管测试使读数复原并稳定后，再用两个试剂空白各进一次样，让仪器取其平均值作为扣底的空白值，随后即可依次测试样。测定完毕后退回主菜单，选择"打印报告"即可将测定结果打印出来。

五、数据处理

如果采用荧光强度测量方式，需先对标准系列的结果进行回归运算（由于测量时"0"管强制为0，故零点值应该输入以占据一个点位），然后根据回归方程求出试剂空白液和试样被测液的砷浓度，再按式（4-4-10）计算试样的砷质量分数或质量浓度：

$$x = \frac{c_1 - c_0}{m} \times \frac{25\,\text{mL}}{1000} \tag{4-4-10}$$

式中　x——试样的砷质量分数或质量浓度，mg/kg或mg/L；

　　　c_1——试样被测液的质量浓度，ng/mL；

　　　c_0——试剂空白液的质量浓度，ng/mL；

　　　m——试样的质量或体积，g或mL。

计算结果保留两位有效数字。

六、思考题

（1）试分析讨论食品中砷的主要来源。

（2）砷对人体健康有哪些影响？

参考文献

[1] 弓爱君，刘杰民，王海鸥.Chemistry of the environment/环境化学［M］.北京：化学工业出版社，2015.

[2] 董德明，花修艺，康春莉.环境化学实验［M］.北京：北京大学出版社，2010.

[3] 顾雪元，毛亮.环境化学实验［M］.南京：南京大学出版社，2012.

[4] 吴翠琴，孙慧，邓红梅.环境综合化学实验教程［M］.北京：北京理工大学出版社，2019.

[5] HJ 479—2009 环境空气　氮氧化物（一氧化氮和二氧化氮）的测定　盐酸萘乙二胺分光光度法.

[6] HJ 504—2009 环境空气　臭氧的测定　靛蓝二磺酸钠分光光度法.

[7] GB/T 7489—1987 水质　溶解氧的测定　碘量法.

[8] GB/T 13193—1991 水质　总有机碳（TOC）的测定　非色散红外线吸收法.

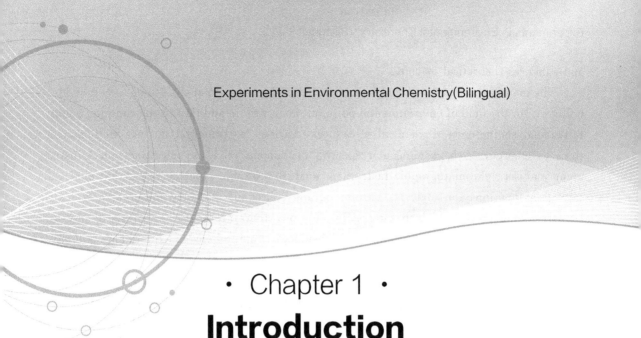

Experiments in Environmental Chemistry(Bilingual)

· Chapter 1 ·
Introduction

In view of the increasing serious problems of environmental pollution and resource shortage, sustainable development and environmental protection have become the theme of the times. The urgent demand for environmental protection has led to a substantial increase in the demand for environmental science talents, and the assessment standards and quality requirements have also increased. At the same time, the training of environmental professionals in the new era is also facing severe tests and challenges.

The important methods and means to solve environmental problems are inseparable from the theoretical knowledge and corresponding professional skills in environmental chemistry, so environmental chemistry plays a very important role in environmental science and engineering and related fields. The experimental link is one of the main ways to cultivate students' innovative spirit and practical ability in environmental chemistry. It plays an irreplaceable role in cultivating innovative talents in environmental science majors.

Section 1 Purpose and task of environmental chemistry experiments

Environmental chemistry experiments cover environmental analysis, environmental monitoring and environmental chemistry. They are based on the theory of chemical experiments and focus on the application of chemical experiments in related majors. The aim is to cultivate students' problem-solving ability (including the mastery and reasonable use of relevant knowledge, the ability to operate and use modern instruments and equipment, the ability to observe, the ability of scientific research and innovation, the ability to independently deal with emergencies, etc.) through practical operation and thinking. It is not a simple verification process of theoretical teaching, but a self-contained course that complements and enhances theoretical teaching.

The main task of environmental chemistry experiments is to analyze and study individual pollutant. In the field of environmental pollutant analysis, in addition to the common quantitative analysis methods of chemical substances, various instrumental analysis methods continue to emerge, such as molecular spectral instrument analysis, electrochemical instrument analysis, chromatographic instrument analysis, atomic spectral instrument analysis, as well as flow injection analysis, chromatography-mass spectrometry and other new analysis and testing means and technologies. In addition, there are also the analysis of physical and chemical properties of environmental substances, comprehensive experiments and environmental monitoring.

Section 2　Laboratory safety rules

Many of the chemicals are flammable, explosive, corrosive and toxic. Therefore, in order to ensure safety, each student is required to attach great importance to safety issues in the mind, and should fully understand the relevant safety knowledge before the experiment. The experiment should be done orderly, strictly as the laboratory safety operating procedures to prevent accidents.

① All bottles containing chemicals should be labeled. Systems for the storage and use of highly toxic chemicals must be formulated and strictly observed. Special counters must be set up for safekeeping with locks. Volatile organic chemicals should be placed in well-ventilated spaces, refrigerators or iron cabinets. Explosive chemicals, such as perchloric acid, perchlorate, hydrogen peroxide and high-pressure gas, should be kept in cool places, not with other flammable materials, and should not be shaken vigorously when moving or starting. The pressure reducing valve of the high-pressure gas cylinder is strictly prohibited from oil contamination.

② It is strictly forbidden to bring food into the laboratory, use experimental utensils as tableware, import reagents (whether it is toxic or not), and eat and smoke in the laboratory. Toxic reagents should not contact the skin and wounds, even enter the mouth. Use pipettes to absorb toxic samples (such as aluminum salt, barium salt, lead salt, arsenide, cyanide, mercury and mercury compounds) and corrosive chemicals (such as strong acid, strong alkali, concentrated ammonia, concentrated hydrogen peroxide, glacial acetic acid, hydrofluoric acid and bromine). Use suction balls for operation, do not use the mouth. Toxic waste liquid is not allowed to be poured into the sewage pipe, should be collected and centralized treatment.

③ Experiments that produce toxic and irritating gases (such as H_2S, Cl_2, Br_2, NO_2, CO, etc.) and use concentrated acids (such as HNO_3, HCl, $HClO_4$, H_2SO_4) or poisons (such as mercury, phosphorus, and arsenide), should be done in the fume hood. When it is necessary to smell the gas, it is strictly forbidden to directly point the nose at the mouth of the bottle or tube. You should gently stir the mouth of the bottle or tube with your hand to blow the gas from the side to yourself, and keep an appropriate distance for sniffing.

④ When opening the volatile reagent bottle (especially in summer), do not make the mouth of the bottle face others or yourself. When opening the bottle mouth, there will be a lot of gas rushing out, which may cause injury accident if you are not careful.

⑤ When using strong corrosive agents such as concentrated acid, concentrated alkali, bromine and lotion, do not splash on skin and clothing. Wear protective goggles and rubber gloves if necessary. When diluting concentrated sulfuric acid, it must be carried out in heat-resistant container. Concentrated sulfuric acid should be slowly poured into the water, and the water should not be poured into concentrated sulfuric acid to avoid splashing. When dissolving heat-generating material such as NaOH or KOH, it must be carried out in a heat-resistant container. If it is necessary to neutralize concentrated acid and concentrated alkali, it must be diluted first.

⑥ When using flammable organic reagents (such as ethanol, acetone, etc.), keep away from the source of fire and close the bottle stopper immediately after use. Potassium, sodium, white phosphorus and other flammable substances in the air should be stored in isolation from the air (potassium and sodium should be stored in kerosene, white phosphorus should be stored in water), and must be taken with tweezers.

⑦ The operation of heating and concentrating the liquid should be very careful, not look down on the liquid being heated, and not point the heated tube mouth to yourself or others to prevent liquid spillage. When concentrating the solution, especially after the appearance of crystals, keep stirring to avoid liquid splashing into eyes or splashing on skin and clothes.

⑧ In the experiment, if flammable chemicals need to be heated or flammable components are removed by heating, they should be carried out slowly on a water bath or electric heating plate. Directly heating with an electric furnace or open flame is strictly prohibited.

⑨ Corrosive articles are strictly prohibited from being baked in the oven.

⑩ The tube should be heated with tube clamp. Hand-held tube heating is not allowed. Glass components heated to red-hot (glass rod, glass tube, beaker, etc.) can not be directly placed on the test table, and must be cooled on asbestos gauge. Since there is no apparent difference between hot glass and cold glass, special care should be taken not to hold the hot end by mistake to avoid scalding.

⑪ For chemical reagents of unknown nature, do not mix arbitrarily. It is strictly forbidden to grind oxidizers with combustibles. It is strictly forbidden to weigh Na_2O_2 or unknown reagents on paper to avoid accidents.

⑫ The cutting of glass tubes (rods), the installation or disassembly of glass instruments, the drilling of plugs and other operations are often easy to cut fingers or hurt the palm, and should be carried out in full accordance with the relevant operating procedures for the safe use of glass instruments. The glass tubes or glass rods should be burnt round immediately after cutting. When installing the rubber tube on the glass tube, the glass tube should be soaked with water or glycerin before the rubber tube covers. Glass debris should be cleaned up in time to prevent accidents.

⑬ All medicines in the laboratory must not be taken outside.

⑭ After the experiment is completed, water, electricity and gas should be turned off, experimental supplies should be arranged, and hands should be washed before leaving the laboratory.

Section 3 Handling of laboratory accidents

In case of any accident during the experiment, the following measures should be taken for rescue.

① If the skin is exposed to a small amount of acid, wash it with plenty of water. If the burn is severe, rinse the skin with saturated $NaHCO_3$ solution after rinsing it with water, then rinse it with water and apply petroleum jelly ointment. If the acid is splashed into the eyes, immediately rinse the eyes with a lot of water, do not direct water to the eyes when rinsing, do not rub the eyes, rinse the eyes with 2% $Na_2B_4O_7$ solution or 3% $NaHCO_3$ solution, and finally rinse the eyes with distilled water. Those with severe burns should be sent to hospital immediately after temporary treatment.

② If the skin is exposed to alkali solution, it can be washed with plenty of water until there is no greasy feeling. It can also be washed with dilute HAc and 2% boric acid solution, then rinsed with water, and coated with boric acid ointment. If the alkali solution is splashed into the eyes, immediately rinse it with plenty of water, then rinse it with 3% H_3BO_3 solution, and finally rinse it with distilled water.

③ If bromine burns occur, the wound can be washed with ethanol or 10% $Na_2S_2O_3$ solution, then rinsed with water, and coated with glycerin.

④ Phosphorus burns can be washed with 5% $CuSO_4$ solution and bandaged with a bandage immersed in $CuSO_4$ solution, or wet compressed with 1:1000 $KMnO_4$, coated with protective agent and bandaged.

⑤ When inhaling irritant or toxic gases such as Cl_2, Br_2, HCl, a small amount of mixed vapor of alcohol and ether can be inhaled to detoxify. When you feel unwell or dizzy due to inhalation of H_2S gas, you should immediately go outside to breathe fresh air.

⑥ If you eat poisons by mistake, you must vomit, wash your stomach and take antidotes again. When vomiting, you could drink a small amount (usually 15-25mL, up to 50mL) of 1% $CuSO_4$ or $ZnSO_4$ solution. After oral administration, put your fingers into your throat to induce vomiting and vomit poisons. Send poisoned person to the hospital immediately for treatment.

⑦ After the burn, you can rinse it with cold water to cool down, use cotton soaked in concentrated (90%-95%) alcohol solution lightly to apply the wound, or wipe the burn with potassium permanganate or picric acid solution, and then apply the burn cream, flower oil or petroleum jelly. If blisters arise, do not break them to prevent infection. People with

Severe scald should be sent to hospital for treatment immediately.

⑧ When you are cut by glass, if there are glass fragments in the wound, they should be picked out first. Then the wound should be washed with disinfection cotton rod or disinfected with iodine, sprinkled with anti-inflammatory powder or applied with anti-inflammatory ointment, and bandaged with wound patches or bandages. If the wound bleeds heavily, the tourniquet should be bandaged on the upper part of the wound to stop bleeding, so as to avoid excessive bleeding and send injuries to the hospital immediately for treatment.

⑨ If there is an electric shock accident, the power supply should be cut off immediately, or the power cord should be pulled away by insulating objects such as wooden sticks. After the electric shock victim is disconnected from the power supply, artificial respiration can be carried out if necessary.

⑩ In case of fire, extinguish the fire immediately, remove flammable chemicals near the fire source, cut off the power supply, and take all possible measures to prevent the spread of the fire. Generally, small fires can be put out with wet cloth, fireproof cloth or sand covered with combustion materials. If the fire is large, appropriate fire extinguishing equipment can be selected according to the cause of the fire. Such as carbon tetrachloride fire extinguisher: suitable for electrical fires, but prohibited for extinguishing CS_2 combustion, otherwise it will produce phosgene and other toxic gases (CS_2 combustion can be extinguished with water, carbon dioxide or foam fire extinguisher); dry powder fire extinguisher: suitable for fighting fires caused by oil, combustible gas, electrical equipment, precision instruments, documents and other items that can be burned with water in the initial stages; carbon dioxide fire extinguisher: suitable for electrical fire extinguishing; foam fire extinguisher: suitable for oil fires, but prohibited in the case of wire or electrical equipment fires.

Note: When oil, wires, electrical equipment, precision instruments, etc. are on fire, it is strictly prohibited to extinguish the fire with water to prevent electric shock or oil drifting with water to expand the burning area. When the clothes on the body are on fire, you should immediately take off the clothes, roll on the spot, or cover the fire with a fireproof cloth. When fighting a fire caused by vapor-toxic chemicals, special attention should be paid to poison prevention.

⑪ When the mercury thermometer is broken or other reasons cause mercury drops, the mercury drops should be immediately gathered together with a brush dipped in water or vaseline. And then the tiny mercury drops are sucked up with a straw or a mercury pickup stick. After that, sprinkle the sulfur powder on the experimental table or ground where mercury is dropped and press hard (so that it generates mercury sulfide), cover it for a period of time before cleaning.

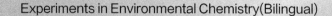

Chapter 2
The usage of common instruments

Section 1 Washing and drying methods of glass instruments

1 Washing of glass instruments

Whether the glassware used in chemical experiments is "clean" often affects the results. Here "clean" means pure. The washing of instruments should be emphasized.

There are many methods for washing instruments, which should be selected according to the requirements of the experiment, the nature of the contamination and the degree of contamination. Generally speaking, the contaminants attached to the instruments are as follows: soluble substances, dust and other insoluble substances, oil contamination and organic matter. The following washing methods can be used respectively.

1.1 Washing with water

It is suitable for washing away only dust and soluble substances, loosely attached solid insoluble substances, oil, and organic matter on the instruments, which can be washed directly with a brush.

1.2 Washing with detergent powder, soap or synthetic detergent

This method can be used for washing instruments with insoluble dirt, oil, and organic matter without precise scale, such as beakers, conical flasks, measuring cylinders, etc. The washing method is to wet the instruments with water (not too much water), sprinkle a little detergent powder or drip a small amount of detergent, and then wash it back and forth with a brush. After the inner and outer walls of the instruments have been carefully scrubbed, the detergent inside and outside the instruments will be washed with tap water until there are no white fine particle powder or foam left. Finally, a small amount of distilled water is used to wash the instruments more than three times, and the calcium, magnesium and chlorine irons brought in from the tap water are washed away.

Note: According to the washing principle of "a small number of times", the amount of distilled water should not be too much each time.

1.3 Washing liquid with chromic acid

Lotion is a saturated solution of potassium dichromate in concentrated sulfuric acid (50g potassium dichromate is added to 1L concentrated sulfuric acid to be heated and dissolved). It has strong oxidation and strong decontantion ability for organic matter and oil. It is suitable for accurate measuring instruments with oil or small tube instruments, such as volumetric flasks, pipettes, burettes and so on. When washing, add a small amount of lotion into the instrument firstly, and then slowly rotate the instrument while tilting until the inner wall of the instrument is fully moistened by the lotion. After a few turns, pour the lotion back into the original bottle, and then wash the residual lotion on the instrument wall with tap water until it is free of the yellow color of chromic acid. Finally, wash the instrument with a small amount of distilled water or deionized water more than three times.

Soak the instrument with lotion for a period of time or use hot lotion will improve the washing efficiency. But you need to pay attention to safety because the lotion is strongly corrosive, do not let the lotion burn the skin. The water absorption of the lotion is strong, and the bottle containing the lotion should be covered after use to prevent water absorption and reduce the decontamination ability. When the lotion is used to appear green (potassium dichromate is reduced to the color of chromium sulfate), it loses its decontamination ability and can no longer be used.

Note: It is preferred to use 1.1 or 1.2 washing methods to clean the instrument. If it is not necessary, chromic acid lotion should not be used because it is expensive and will cause serious pollution.

In addition, the glass instruments can be cleaned with an ultrasonic cleaner. The standard for testing whether the glass instrument is clean is that the instrument is transparent and the inner wall forms a uniform water film, and the water does not gather into water droplets when you turn the instrument upside down.

The instrument shall not be wiped with cloth or paper after washing, otherwise the fiber of the cloth or paper will remain on the wall of the instrument and stain the instrument.

2 Drying of glass instruments

In addition to being washed, some instruments require being dry without water film. There are several drying methods:

① Air dry: The instruments that need to be air-dried can be placed in the cabinet or on the rack after being washed and be dried naturally.

② Blow dry: The instruments are directly dried by hot air of hairdryer.

③ Heat dry: Instruments that can be directly heated, such as test tubes, beakers and evaporating dishes, can be directly dried in a small fire on a gas stove or an alcohol lamp. When baking the test tube, we should pay attention to: the test tube mouth should be downward, so as not to reverse the flow of water beads and burst the test tube; you should continue to move the test tube back and forth while baking; after baking until no water beads, then turn the tube mouth up to get the water vapor out.

④ Bake dry: The washed glass instruments (excluding the high-precision capacity instruments) can be placed in the electric oven and dried at about 150℃. Before the instruments are put into the oven, the water should be poured as far as possible, and an enamel tray should be placed on the bottom layer of the oven to prevent the water drops from falling on the electric heating wire and damaging the oven.

⑤ Using acetone, ethyl alcohol and other organic solvents to quickly dry. When using organic solvents such as acetone and ethanol to quickly dry measuring instruments with scales (such as pipettes, measuring cylinders, volumetric bottles, etc.), heating can not be used to dry because it will affect the precision of the instruments. You can use volatile organic solvents (most commonly ethanol or 1 : 1 mixture of ethanol and acetone) to wash the washed glass instruments. The water on the wall of the instruments and these organic solvents dissolve and mix with each other. Pour out the water-containing mixture (water-containing mixture can be recovered) and a small amount of residual liquid can volatilize in a short time, leaving the instruments dry naturally.

Section 2 Water for environmental chemistry experiments and its preparation

In the laboratory, a lot of water is needed for cleaning instruments, preparing solutions, and analyzing and determining, but tap water contains various impurities and does not meet the requirements of the experiments. Pure water is usually obtained by distillation or ion exchange.

1 Distilled water

The water produced by condensation of natural water through a still is called distilled water. Distilled water in the laboratory is prepared by distillation flask, condensing tube and other instruments. Tap water is heated and vaporized in the flask, and the water vapor rises and condenses through the condensing tube and flows into the receiver for collection, as shown in Fig. 2-2-1. Electric water distillers are also available for laboratory use. Distilled water still contains a trace of impurities, and its conductivity is 2.8×10^{-6} S/cm at 25℃.

2 Deionized water

Deionized water is pure water prepared by ion exchange method. This method has been widely used in laboratories and industrial departments. In addition to the preparation of dei-

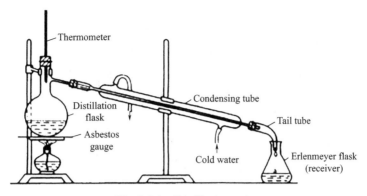

Fig. 2-2-1 Distillation unit

onized water, ion exchange method can also be used for the separation and purification of rare metals, the recovery of metals, the extraction of antibiotics, and other aspects.

2.1 Exchange principle

The ion exchange method is a method to remove impurity ions in water by passing tap water through an ion exchange column equipped with cation or anion exchange resin in turn.

Ion exchange resin is solid polymer with reticulated skeleton structure. The active functional groups in its skeleton structure exchange with ions in water. For example, polystyrene sulfonic acid cation exchange resin is a copolymer of styrene and a certain amount of divinylbenzene. After concentrated sulfuric acid treatment, sulfonic acid groups ($-SO_3H$) are introduced into the benzene rings of the copolymer. It is strong acidic cation exchange resin. When the resin is immersed in water and fully expanded, the space in the skeleton enlarges. Hydrogen ions of sulfonic acid groups ($-SO_3H$) on the benzene rings can exchange with cations in water (such as Ca^{2+}, Na^+, Mg^{2+}, etc.). The reaction is as follows:

$$R-SO_3^- H^+ + Na^+ \rightleftharpoons R-SO_3^- Na^+ + H^+$$
$$2R-SO_3^- H^+ + Ca^{2+} \rightleftharpoons (R-SO_3^-)_2 Ca^{2+} + 2H^+$$

R is a copolymer of styrene and divinylbenzene.

Anion exchange resin introduce alkaline groups such as amino groups into the network framework of copolymers, such as quaternary ammonium salt strong alkaline anion exchange resin $R\equiv N^+ OH^-$, in which OH^- can be exchanged with anion X^- in water. The reaction is as follows:

$$R\equiv N^+ OH^- + X^- \rightleftharpoons R\equiv N^+ X^- + OH^-$$

2.2 Exchange device

Ion exchange columns are usually used to prepare purified water, and the specific connections are shown in Fig. 2-2-2. The tap water first passes through the cation exchange column to remove Ca^{2+}, Mg^{2+} and other cations in the water, and then flows into the anion exchange column, where the anion in the water is exchanged with the OH^- in the exchange resin and the H^+ and OH^- replaced in the two columns are combined to form H_2O. Deionized water can be obtained after several exchanges.

Experiments in Environmental Chemistry (Bilingual)

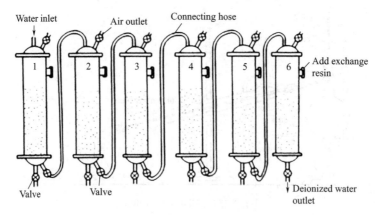

Fig. 2-2-2 Schematic diagram of deionized water preparation device
1, 4—Cation exchange column; 2, 3, 5, 6—Anion exchange column

2.3 Preparative operation

Some oligomers, pigments, limestone and so on are often found in the newly purchased ion exchange resin. Impurities must be removed, and resin must be transformed while using the instrument.

(1) Rinsing treatment

The new resin is placed in a basin, washed repeatedly with water below 40℃ until the supernatant to be colorless. Then the anion and cation exchange resin are respectively loaded into the exchange columns, soaked in water for 24h.

(2) Transformation treatment of strong acidic cation exchange resin

Cation exchange columns are connected in series, and 2mol/L HCl is added to replace the water in the columns. The columns are stationary for 2-3h, and the pH of the effluent is 3-4 after washing with water. Add 2mol/L NaOH solution and stand for 2-3h. Then wash with water to the pH of the effluent is 9-10. Finally, 2mol/L HCl is used for transformation treatment, and 1kg resin consums about 4.5L acid.

Note: The flow rate of HCl should not be too fast during the transformation, and it should be about 60mL/min. After adding hydrochloric acid, wash columns with deionized water until the pH of the effluent is about 4.

(3) Transformation treatment of strong basic anion exchange resin

Anion exchange columns are connected in series (the dosage of anion exchange resin should be twice that of cation exchange resin). Water in columns will be replaced by 2mol/L NaOH solution, and columns will be stationary for 2-3 hours and then be washed with deionized water (or water through cation resin) until the pH of effluent is 9-10. After washing with 2mol/L HCl solution, the pH of the eluent is 3-4, and then transform columns with 2mol/L NaOH solution. The calculation of alkali consumption and the outflow rate of alkali solution are the same as those of the cation exchange resin. After that, rinse columns with deionized water until the pH of the eluent is 9-10.

The above activated and transformed ion exchange resin is reassembled into columns, and the air bubbles in the resin layer are removed. Then the incoming water can be treated, and the initial outflow water should be discarded.

Ion exchange resin will lose its exchange capacity after a period of time. At this time, regeneration should be carried out. The treatment methods are as follows: cation exchange column is treated with HCl solution, and anion exchange column is treated with NaOH solution. The concentration, dosage and treatment method of solution are the same as those of resin transformation.

2.4 Quality inspection of deionized water

Acidity and alkalinity: Deionized water is added into two test tubes, methyl red indicator is added into one test tube and bromothymol blue indicator is added into the other test tube. The liquid doesn't show red or blue.

Calcium and magnesium ions: Take 10mL of deionized water to be checked, add ammonia buffer solution (add 20g NH_4Cl to 1L 2mol/L $NH_3 \cdot H_2O$ solution), adjust the pH to about 10, add chrome black T indicator, and the water does not show red.

Chloride ions: Add deionized water into test tube, acidify liquid with 2mol/L HNO_3, then add 1 drop of 0.1mol/L $AgNO_3$ solution, shake well, and observe no turbidity.

Conductivity: Conductivity of deionized water is measured by the conductivity meter to assess the quality of water. The higher purity of water, the fewer impurity ions, the higher resistivity and the lower the conductivity of water. The conductivity of deionized water is 8.0×10^{-7}-4.0×10^{-5} $S \cdot cm^{-1}$ at 25℃.

Section 3 Common instruments for environmental chemistry experiments

1 Balance

1.1 The accuracy of weighing for various balances

Different levels of electronic balances are used in the laboratory. The accuracy of the mass of an object measured by various balances is also different.

One ten-thousandth balance generally can be said to be accurate to 0.1mg, that is, 0.1mg is an inaccurate number.

One thousandth balance generally can be said to be accurate to 1mg, that is, 1mg is an inaccurate number.

One hundredth balance can be said to be accurate to 10mg, that is, 10mg is an inaccurate number.

One tenth balance generally can be said to be accurate to 0.1g, that is, 0.1g is an inaccurate number.

1.2 Electronic balance

The latest generation of balance is electronic. It uses electronic devices to complete the regulation of electromagnetic force compensation and achieves the balance of force in gravity field. It can also achieve the balance of moment in gravity field through the regulation of electromagnetic moment. The structure of common electronic balances is electromechanical, consisting of load acceptance and transfer device, measurement and compensation device, etc. Balances can be divided into two types: top load-bearing and bottom load-bearing. At present, most of them are top load-bearing balances. There are two kinds of calibration methods: internal calibration and external calibration. The former is the standard weight is pre-installed in the balance. After starting the calibration key, the balance can automatically add code for calibration. The latter needs to be corrected manually by putting the standard weight on the weighing plate. Metler-Toledo L-IC series balance is shown in Fig. 2-3-1.

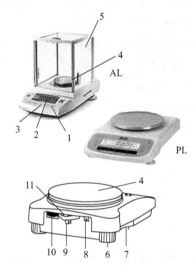

Fig. 2-3-1 Structure of balance

1—Operation key; 2—Display screen; 3—Model plate with the following parameters: "Max" means the biggest weighing, "d" means actual indexing value; 4—Weighing plate; 5—Windshield; 6—Level adjusting screw; 7—Scale hook hole for hanging down weighing; 8—AC power adapter socket; 9—RS232c interface; 10—Anti-theft lock link ring; 11—Level bubbles

1.3 Calibration of balance

In order to obtain accurate weighing results, calibration must be carried out to adapt to local gravity acceleration. Calibration is necessary in the following circumstances: before the first use of balance weighing; during the weighing work; after changing the placement position. Two calibration methods are as follows:

(1) Internal calibration

Keep the weighing plate empty and hold the "CAL" key until the "CAL Int" appears on the display screen and then loosen the key. The balance is automatically calibrated. When the information "CAL Done" appears on the display screen for a short time and then "0.00g" appears, the balance calibration ends. The balance returns to the weighing mode of work and waits for weighing. (Fig. 2-3-2) (Note: Please make sure that the CAL item in MENU is CAL Int.)

(2) External calibration

Prepare the calibration weights. Let the weighing plate be empty. Press and hold the "CAL" key until the "CAL" appears on the screen and release it. The values of required calibration weights will flicker on the display screen. Place the calibration weights (in the center of the weighing plate). The balance is automatically calibrated. When "0.00g" flickers, remove the weights. Calibration of the balance is completed when the information "CAL Done" appears on the display screen for a short time, followed by another "0.00g". The balance returns to the way of weighing and waits for weighing. (Fig. 2-3-3) (Note: Please confirm that CAL item in MENU is CAL.)

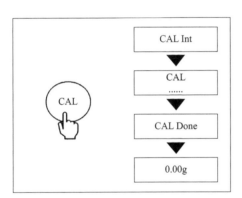

Fig. 2-3-2　Internal calibration of balance

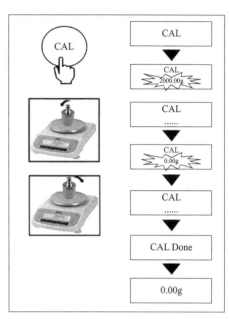

Fig. 2-3-3　External calibration of balance

1.4　Weighing

(1) Power on and off

Power on: Check the level meter before turning on the balance. If not, ask the teacher to adjust it to be level by adjusting the level adjusting screw. Turn on the power supply and preheat for 20 minutes (if the balance hasn't been used for a long time), then the display can be turned on for operation.

Leave the weighing plate unloaded and click "ON" key. The balance is self-checked for display (all fields on the display screen are lighted up for a short time). When the balance returns to zero, it can be weighed.

Power off: Press the "OFF" key until "OFF" appears on the screen, and then loosen the key. (Fig. 2-3-4)

(2) Simple weighing

Place the weighing sample on the plate. Wait until the steady state probe disappears. Read the weighing result. (Fig. 2-3-5)

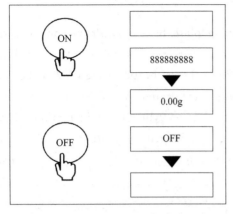

Fig. 2-3-4 Power on and off of the balance

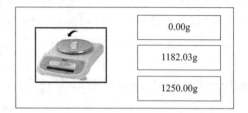

Fig. 2-3-5 Simple weighing of balance

(3) Quick weighing (decrease reading accuracy)

The balance allows for a reduction in reading accuracy (digits after decimal points) to speed up the weighing process. The balance generally works at normal accuracy and speed. When pressing the "1/10d" key, the balance works at a lower reading accuracy (one less decimal point), but it can display the results more quickly. After pressing the "1/10d" key, the balance returns to its normal reading accuracy. (Fig. 2-3-6)

(4) Peeling

Place the empty container on the balance. Screen display the weight. Press the "→O/T←" key. When the container is filled with weighing samples, the weight is shown on the screen. If the container is removed from the balance, the leather weight will be negative. Leather weight will remain until you press the "→O/T←" key again or turn off the balance. (Fig. 2-3-7)

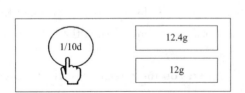

Fig. 2-3-6 Rapid weighing of balance

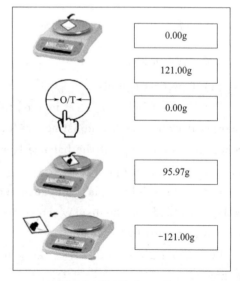

Fig. 2-3-7 Scale peeling

(5) Notes

① Before opening the balance, check whether the balance's weighing plate and its interior are clean. Before the balance is opened, clean the balance with a brush.

② Designated person is responsible for the starting up, preheating, and calibration of the balance. Users only need to press "ON", "CAL", and "OFF" keys when weighing, and other keys are not allowed to press arbitrarily.

③ The base of the balance is relatively light, resulting in insufficient stability, and easy to be bumped and displaced, resulting in poor level and affecting the weighing results. Therefore, special attention should be paid to the use with light and slow actions. It's necessary to check the level meter frequently.

2 3200 Atomic absorption spectrophotometer

2.1 Principle

Atomic absorption spectrophotometry (AAS) is a method for quantitative analysis of the absorption of specific spectral lines by atomic vapor produced by substances. Like spectrophotometry, according to Lambert-Beer's law, the relationship between absorbance and the number of ground state atoms in the sample is as follows:

$$A = \lg \frac{I_0}{I} = K'LN \tag{2-3-1}$$

Where, A——absorbance;

I_0——the intensity of incident light;

I——the intensity of light passing through atoms;

K'——constants;

L——the thickness of vapor passing through atoms, cm;

N——The number of ground state atoms.

Because the flame temperature of AAS is less than 3000K, most of the atoms in the flame are ground state atoms, and the excited state atoms are very few. The number of ground state atoms is proportional to the concentration, and the flame width is constant. Therefore, the relationship between the absorbance and the concentration of the tested elements can be expressed as follows:

$$A = KC \tag{2-3-2}$$

Where, A——absorbance;

K——constants;

C——the concentration of the tested elements, mol/L.

Eq. (2-3-2) is the basic formula for quantitative analysis.

The general devices of atomic absorption spectrophotometer mainly include light source, atomizer, spectroscopic system, detection, and display system.

(1) Light source

At present, hollow cathode lamp is mainly used to provide sharp-line light source. The cathode is made of the metal of the element under test into a hollow shape and sealed in a cylindrical glass tube at low pressure, which is filled with inert gas of 2mm mercury column. The front end of the glass tube is a quartz window. (Fig. 2-3-8)

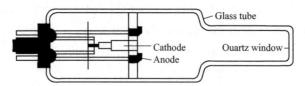

Fig. 2-3-8　Hollow cathode lamp

(2) Atomizer

Atomizers are divided into flame atomizers and non-flame atomizers. The flame atomizer is composed of a sprayer, a spray chamber, and a burner, which is used to convert the measured elements in the sample to be tested into ground atoms that are uncombined, non-excited, non-ionized, and unassociated. The most widely used non-flame atomizer is the graphite furnace atomizer, which is composed of graphite tube (cup), furnace body (protective gas system), and power supply. The element to be measured in the sample is atomized by electric heating, cathode sputtering, high-frequency induction or laser. Its sensitivity is several orders of magnitude higher than the flame method. The absolute sensitivity can reach 10-12g. The sample amount is small, generally only 1-100μL, but the determination precision is poor.

(3) Spectroscopic system

There are still some interferences in atomic absorption spectroscopy, especially the multi-line elements. Therefore, the resonance absorption line is separated from other interfering non-absorption lines by the spectrometer. At present, grating spectroscopy is often used.

(4) Detection and display system

It consists of photomultiplier tube and amplifier. Firstly, the optical signal is converted into electrical signal. Then the measured signal is amplified, while the non-measured signal is removed. Finally, the results display on the digital display window or recorder.

2.2　Analytical methods

(1) Standard curve method

Prepare a series of standard solutions with different concentrations. Under the selected conditions, adjust the absorbance of the blank solution to zero, then measure the absorbance of each standard solution, and draw the standard curve of absorbance versus concentration. Samples must also be measured under the same conditions. The unknown concentration can be found from the working curve.

(2) Accession method

The absorbance of the sample solution is measured firstly. Then a certain amount of standard solution similar to the sample solution is added to the sample solution. Finally, the absorbance at this time is measured. Then:

$$A_x = KC_x \qquad (2\text{-}3\text{-}3)$$
$$A_0 = K(C_0 + C_x) \qquad (2\text{-}3\text{-}4)$$

That is:

$$C_x = C_0 \times \frac{A_x}{A_0 - A_x} \tag{2-3-5}$$

Where, A_x——the measured absorbance of the sample solution;

A_0——the corresponding absorbance of standard solution;

C_x——the concentration of the sample solution, mol/L;

C_0——the concentration of standard solution, mol/L;

K——constants.

The unknown concentration can be calculated, and the matrix effect can be eliminated by this method.

(3) Extrapolation method

Several sample solutions of the same volume are added with standard solutions of different proportions from the second part, and then diluted to a certain volume. The concentration of each solution will be C_x, $C_x + C_0$, $C_x + 2C_0$, $C_x + 3C_0$, and $C_x + 4C_0$, respectively. The corresponding absorbance is A_x, A_1, A_2, A_3, and A_4, respectively. Draw the curve of A versus C. The intersection of extrapolated straight line and concentration axis is the concentration of the certain element in the original sample. (Fig. 2-3-9)

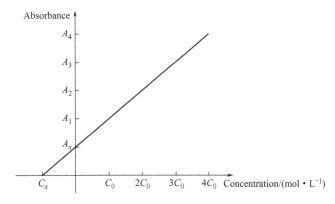

Fig. 2-3-9   Extrapolation working curve

2.3 Operation procedure

(1) Instrument operating steps

① Strictly check the external wiring of the circuit before switching on the instrument, and release all keys on the electronic control panel (Fig. 2-3-10).

② Open the upper left door of the instrument, insert the hollow cathode lamp of the element to be measured into the lamp holder, place it on the lamp stand and press it with spring, and make the lamp cathode roughly equal to the mark on the lamp stand.

③ According to Fig. 2-3-10, press the main power switch (12) and the lamp power switch (39). Adjust the lamp current adjusting knob (37) after the hollow cathode lamp emits light, so that the current of the hollow cathode lamp is at the selected value.

④ The slit knob is placed in the second gear. According to Fig. 2-3-10, select the wavelength scanning button (13, 14) and the wavelength conversion button (17). Look at the spectral line of the element of the lamp from the wavelength display window (18), loosen

Experiments in Environmental Chemistry (Bilingual)

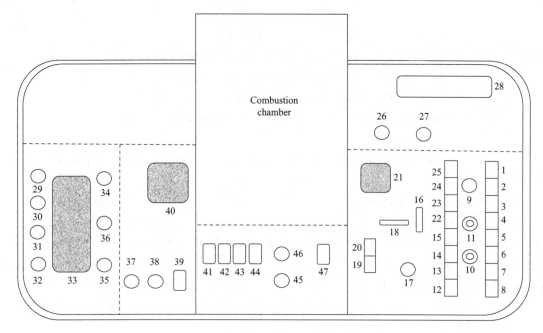

Fig. 2-3-10 Panel diagram of 3200 AAS

1—1 Second button; 2—3 Seconds button; 3—8 Seconds button; 4—15 Seconds button; 5—Background correction button; 6—Curve aligning button; 7—Flame emission button; 8—Condition setting; 9—Aligning button; 10—Gain knob; 11—Range expansion; 12—Main power switch; 13—Wavelength scanning button (↓); 14—Wavelength scanning button (↑); 15—Reference energy button; 16—Wavelength manual wheel; 17—wavelength conversion button; 18—Wavelength display window; 19—Reading button; 20—Zero button; 21—Energy meter; 22—Direct concentration button; 23—Automatic gain button; 24—Peak value holding button; 25—Integral button; 26—Spectral slit knob; 27—Damping knob; 28—Digital display window; 29—Pressure meter; 30—Atomizer knob; 31—Oxidant gas knob; 32—Knob of oxidant gas pressure stabilizing valve; 33—Rotameter; 34—Acetylene meter; 35—Knob of acetylene pressure stabilizing valve; 36—Acetylene regulating knob; 37—Lamp current adjusting knob; 38—Pull handle; 39—Lamp power switch; 40—Ammeter; 41—Gas source switch; 42—Air-nitrous gas switch; 43—Oxidant gas switch; 44—Acetylene gas switch; 45—Burner fore-aft adjusting knob; 46—Burner height adjusting knob; 47—Manual ignition switch

the scanning button, push the conversion button into the middle space, slowly move the wavelength manual wheel (16) around the most sensitive line of the element by hand, and observe the energy meter (21). In the process of switching the wavelength manual wheel, there are generally two situations as follows.

a. The energy meter does not reflect (or in the white area). Adjust the gain knob (10), and then move the wavelength manual wheel back, so that the energy meter's pointer is in the blue area.

b. The energy meter's pointer goes beyond the blue area. Adjusts the gain knob (reduces the negative high pressure) so that the energy meter's pointer returns to the blue area, and moves the manual wheel to maximize the energy and make pointer to the blue area.

Note: When moving the wavelength manual wheel, the reading number of wave length is from long wave to short wave. If the automatic gain button (23) is pressed, the wavelength cannot be adjusted, and components will be damaged in serious cases.

⑤ Rotate the position of the hollow cathode lamp to maximize the indication of energy meter (do not easily adjust threaded sleeves and adjusting screws on element lamp holder without special circumstances).

⑥ Press the zero button (20), and the digital display automatically returns to zero. Connect the power switch of the recorder, adjust the zero knob on the recorder, put down the recorder pen, and select the speed of paper to make the recorder work normally.

⑦ Adjust the burner fore-aft adjusting knob and height adjusting knob (45, 46) and burner lamp holder and handle, so that the burner lamp holder gap and optical axis are parallel. Correct the burner light in the appropriate position below the optical axis (about 3-6mm).

⑧ Water is injected into the premixing chamber to ensure that the waste liquid discharge pipe is filled with water, so that air leakage can be avoided. Water in the waste liquid discharge pipe should be maintained throughout the work. Open the suction hood switch, start the air compressor, and then open the acetylene cylinder.

⑨ Turn on the gas source switch (41) on the gas circuit control board and regulate the pressure stabilizing valve to make air pressure is 196kPa. The pressure of acetylene is 59kPa by adjusting the acetylene pressure stabilizing valve. Connect the acetylene switch and adjust the acetylene needle valve, so that the flow is about 1.5L/min.

⑩ After acetylene and oxidant gas flow for about 5 seconds, the manual ignition switch (47) is pressed and the flame on the burner lamp is ignited, and then switch can be released.

Note: When extinguishing fire, turn off acetylene firstly, then turn off oxidant gas.

⑪ After the flame is ignited, the gas flow rate is adjusted to the required flame state and the burner is preheated.

⑫ Spray the blank deionized water later and continue to preheat the burner. After that, the blank deionized water spray should not be interrupted for a long time.

⑬ Carefully select the analytical conditions (see the instrument operating conditions selection), and complete the determination of standard samples and samples in turn.

⑭ At the end of the work, spray with deionized water. Then cut off the burning gas source, burn out the remaining acetylene in the pipeline, and then cut off the power supply of the air compressor after the flame goes out.

(2) Instrument operating conditions selection

① Lamp current selection: Fixed conditions include wavelength, slit width, burner height, air flow rate, acetylene flow rate, test liquid lifting amount. By changing the current of the hollow cathode lamp, the corresponding absorbance A is obtained, so as to find out the best working current of the lamp.

② Selection of burner height: Fix the above conditions, only change the burner height, measure the change in absorbance A of the element, so as to find the optimum burner height.

③ Selection of flow ratio of oxidant gas to combustion gas: Fix the above conditions, and make the flow rate of acetylene only change between 0.5 and 2.0L/min. The change of elemental absorbance A is measured to find out the best flow ratio of acetylene to oxidant gas.

④ Selection of slit width: Fix other conditions, just change the width of the slit, measure the change of elemental absorbance A, and find the best position of the slit.

3 2100P Portable turbidimeter

3.1 Calibration of turbidimeter

For the best accuracy, use the same sample pool or four matching sample pools throughout the calibration period. Sample pools should normally be inserted in the direction labeled during matching.

① Rinse the clean sample pool several times with diluted water. Then diluted water or StablCal < 0.1 NTU standard solution is added to the sample pool to the scale line (about 15mL).

Note: The same diluted water as which is prepared for standard liquid must be used in this step.

② Put the sample pool into the sample pool box of the instrument, and make the direction mark on the sample pool is in line with the direction mark in front of the sample pool box. Cover the pool and press the "I/O" key.

Note: Before pressing the "CAL" key, select the signal averaging mode option (on or off). In calibration mode, the signal averaging function does not work.

③ Press the "CAL" key, and the "CAL" and "S0" icons will be displayed on the screen ("0" will flicker). The 4-bit displayed value shows the previously calibrated value of S0 standard liquid. If the blank value is forcibly assigned to 0.0, the display will be blank. Press "→" key to get a digital display value.

④ Press "READ" key, and the instrument will count from 60 to 0 (67 to 0 if the signal average function is turned on). Read the blank value and make it as a correction factor for calculating the test value of 20 NTU standard liquid. If the turbidity value of diluted water is greater than 0.5 NTU, the word "E1" will appears on the screen when calculating calibration. The display will automatically progress to the next standard liquid test. Take the sample pool out from the sample pool box.

Note: By pressing the "→" key instead of reading the value of diluted water, the turbidity of the diluted water can be forcibly assigned to zero. The display screen will display "S0 NTU". To continue test the next standard solution, the "↑" key must be pressed.

⑤ The screen will display "S1" icon ("1" will flicker) and "20 NTU" or the valve of previously calibrated S1 standard liquid. If the value is incorrect, press the "→" key until the number you need to edit flickers, and then edit the value. Use the "↑" key to scroll to the correct number. After editing, the mixed standard solution of 20NTU StablCal or 20NTU Foamazin is added into the clean sample pool to the marking line. The sample pool is put into the sample pool box of the instrument, and the direction mark on the sample pool is aligned with the direction mark in front of the sample pool box, and the pool is covered.

⑥ Press "READ" key, and the instrument will count from 60 to 0 (67 to 0 if the signal averaging function is turned on), then test turbidity and store the value. The display will automatically progress to the next standard liquid test. Take the sample pool out from the sample pool box.

⑦ The screen will display the "S2" icon ("2" will flicker) and "100 NTU" or the value of previously calibrated S2 standard liquid. If the displayed value is incorrect, press the "→" key until the number you need to edit flickers, and then edit the value. Use the "↑" key to scroll to the correct number. After editing, add 100NTU StablCal standard solution or 100NTU Foamazin standard solution, which is fully mixed, into the clean sample pool to the marking line. Put the sample pool into the sample pool box of the instrument, and make the direction mark on the sample pool align with the direction mark in front of the sample pool box. Cover the pool.

⑧ Press "READ" key, and the instrument will count from 60 to 0 (67 to 0 if the signal averaging function is turned on), then test turbidity and store the value. The display will automatically progress to the next standard liquid test. Take the sample pool out from the sample pool box.

⑨ "S3" icon ("3" will flicker) and 800NTU or the value of previously calibrated S3 standard liquid will be displayed on the screen. If the value is incorrect, press the "→" key until the number you need to edit flickers, and then edit the value. Use the "↑" key to scroll to the correct number. After editing, add fully mixed 800NTU StablCal or 800 NTU Foamazin standard solution into the clean sample pool to the marking line. Put the sample pool into the sample pool box of the instrument, and make the direction mark on the sample pool align with the direction mark in front of the sample pool box. Cover the pool.

⑩ Press "READ" key, and the equipment will count from 60 to 0 (67 to 0 if the signal averaging function is turned on), then test turbidity and store the value. The display will automatically progress to the next standard liquid test. Take the sample pool out from the sample pool box.

⑪ Press the "CAL" key to confirm the calibration value. The instrument will automatically return to the test mode.

Note: Press the "CAL" key to complete the calibration of calibration coefficient. If a calibration error occurs during the calibration process, an error message will appear when the "CAL" key is pressed.

Matters needing attention are as follows.

If "I/O" key is pressed during calibration, new calibration data will be lost, but old calibration data will be used for testing. Once in the calibration mode, only "READ", "I/O", "↑", and "→" keys work. Signal average function and range selection mode must be selected before entering calibration mode.

If "E1" or "E2" appears on the screen, it indicates that an error occurred during the calibration process. Check the preparation and calibration process of the standard solution, and recalibrate if necessary. Press the "DIAG" key to clear the error message ("E1" or "E2"). If you want to continue testing without recalibration, press the "I/O" key twice to restore to the original calibration value. "CAL?" showing indicates that an error occurs during the calibration process. The original calibration value may not be restored. You can choose to either recalibrate or use the current calibration value. If "CAL?" flickers, the instrument is using the default calibration value.

To view the calibration value, press the "CAL" key and then press the "↑" key to see the value of the calibration standard solution. As long as the "READ" key is not pressed and the "CAL" does not flicker, the calibration value will not be updated. Pressing the "CAL" key again will return to the test mode.

3.2 Turbidity test procedures

① Use a clean container to collect representative samples. Add the sample into the sample pool to the calibration line (about 15mL). Carefully hold the top of the sample pool during operation. Then cover the sample pool.

Note: if no button is pressed within 5.5minutes, the device will be automatically turned off. Press the "I/O" key to restart the device.

② Wipe the sample pool with a soft and lint free cloth to remove water droplets and fingerprints.

③ Add a small drop of silicone oil and wipe it with an oil cloth, so that the whole surface of the sample pool is evenly distributed with a layer of silicone oil.

④ Press the "I/O" key and then the instrument will be opened. Please place the instrument on a flat and stable surface. When testing, do not hold the instrument by hand.

⑤ Put the sample pool into the sample pool box of the instrument. Make sure that the diamond mark or direction mark is aligned with the direction mark raises in front of the sample pool box. Cover it with cover plate.

⑥ Press the "RANGE" key to select the manual or automatic selection range mode. When the equipment is in the mode of automatic selection range, the display screen will display "AUTO RNG".

⑦ Press "SIGNAL AVG" key to select the appropriate signal averaging mode. "SIG AVG" will be displayed on the screen when the instrument uses signal averaging mode. If the sample causes a noise signal (i.e. the display value changes constantly), use the signal averaging mode.

⑧ Press "READ" key and the "----NTU" will be displayed on the screen, and then the turbidity value in NTU will be displayed. Record the turbidity value after the light signal is turned off.

Note: The instrument's default will be the last selected mode of operation. If the automatic selection range and signal averaging mode are chosen in the previous test, these options would be automatically selected in the next test.

The 2100P portable turbidimeter has been calibrated with Formazin first-class standard solution in the factory, so no further calibration is required before use. Hash recommends that Formazin should be used every three months for recalibration or the number of calibrations should be increased based on experience. The Gelex secondary standard liquid attached to the instrument has indicated the basic range of use, but after Formazin calibration, its value must be determined again before use.

4 FB2200 Acidometer

4.1 Calibration of acidmeter

① Click the "Exit" button to power on and enter the self-test program;

② Place the cleaned electrode into the standard buffer with the pH of 6.86, stand for 30s, press the "Calibration" button to start calibration, and the screen displays "$\sqrt{Auto}$". The first point calibration is completed;

③ Take out the electrode, clean it, select a buffer solution with the pH of 4.01, and complete the second point calibration according to the above method;

④ After calibration, the instrument automatically switches to the calibration result interface. The "☺" displayed in the upper left corner of the display indicates that the electrode has good performance and can be used normally;

⑤ Clean the electrode with deionized water and blot dry with filter paper, put the electrode into a beaker with the liquid to be measured, stir, and stand for 30 seconds. Click "Reading" to start measuring;

⑥ The instrument emits a beep and the screen displays "$\sqrt{Auto}$";

⑦ Stand for about 10s, and the value displayed on the screen is the final measured value. Record the data;

⑧ Take out the electrode and clean it.

Two sets of two-point calibration schemes of 6.86/4.01 and 6.86/9.18 are implemented in the software design of acidmeter for application according to requirements within the range of 0-14.00pH.

4.2 PH measurement

Before the measurement, the standard liquids of 6.86 and 4.01 can be measured back to check whether the measuring accuracy of the instrument is within the required range. If there is any discrepancy, the above calibration steps can be repeated, and the equipment can be calibrated twice. Single point repeated calibration can be carried out for zero or slope (the reason why the primary calibration is not in place may be that the electrode is affected by temperature factors).

After the calibration operation and verification steps are completed, the cleaned electrodes can be placed in the solution under test. Stir sufficiently to make the induction bulb of the pH sensor fully contact with the solution under test. When the value on the display screen is stable, the value can be read.

Suggestions: The process from calibration to measurement should be carried out with the cooperation of magnetic stirrer as far as possible. If the measured solution contains organic substances or factors that interfere with the activity of hydrogen ions, proper treatment should be carried out firstly to avoid instability in the measurement results of the instrument. In order to ensure the objectivity of the measurement, from calibration of the instrument to measurement, the water used for electrode cleaning must be deionized water or distilled water.

The instrument has the function of remembering the current calibration value. When it is used frequently without changing the electrode and the calibration value, the calibration operation can be omitted every time when the instrument is used repeatedly. The premise is that the power supply of the instrument can be turned off, but the power supply cannot be cut off.

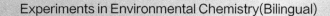

Experiments in Environmental Chemistry(Bilingual)

Chapter 3
Pretreatment technology of environmental samples

Section 1 Collection methods of gas samples

1 Collection of aerosol(smog) samples

Aerosols in the environment have particles ranging from 0.01 to 10.0 μm, sometimes even larger. The composition of these aerosols is usually related to particle size. Batch sampling usually collects aerosols by filtration, collision or electrostatic adsorption. The main basis of the selection method is to avoid the transformation of gas particles on the filter membrane, to avoid the disappearance of samples through aerodynamic action, and to avoid the contamination caused by the filter material. Filtering is the most commonly used method, and high-speed samplers (gas flow is 0.1-3m^3/h) can be used. The selection of filter membranes is very important. Particle size, collection efficiency, and possible contamination need to be considered. Plastic fibre filter paper has poor mechanical strength. It is preferable to use fibre filter paper and quartz filter paper, but the applicability of membrane filtration is more common. This technology has good collection efficiency and low background of trace elements, but has high air flow resistance.

There are some problems in collecting mercury, organic mercury, and other volatile elements such as lead, arsenic and selenium in the atmosphere by using the filter membrane sampler. The main reason is that the working conditions on the surface of the filter membrane are easy to lead to desorption and volatilization of the collected samples. Both high-speed and low-speed sampling devices face this problem. For high-speed samplers, the surface velocity is similar to that of low-speed samplers because of the need for larger filter paper.

Some commercial samplers usually collect atmospheric particles by collision. These instruments have the ability to collect aerosols according to different size and components. Electrostatic adsorption is rarely used in analytical chemistry, but it has a good collection function. There are many commercial instruments used in occupational health surveys to collect certain particles in the atmosphere. Most of them are filter type and can be carried with people. At present, the study of trace metals in the atmosphere mainly focuses on the chemical behavior of mercury and lead. It is very important to study the presence and chemical behavior of organic lead compounds in the atmosphere. Although the amount of lead used as an explosion-proof agent for gasoline has been significantly reduced in general, there are some emissions or radiations from smelters. Organic lead is still considerable. Organic mercury widely exists in the atmosphere due to natural and artificial emissions. Its harmful chemical properties have become the focus of research. For these two elements, chemical speciation analysis not only depends on their presence and reaction pathways in the atmosphere, but also on the toxicity of inhalation. These two elements can be attached to particulates as volatile gas or exist as discrete particulates themselves. Volatile organic lead, including tetraethyl lead, tetramethyl lead, and their degradation products, is about 1% to 4% of the total lead in the city. In the combustion of lead-containing materials, the presence of dichloroethylene or dibromoethylene causes most lead compounds to be discharged in the form of chloride or bromide. Under the action of sulfate, these halides are transformed into $PbSO_4$. In the smelting process, the emissions contains Pb, PbO, PbS, $PbSO_4$ and so on.

In the surrounding atmosphere, elemental mercury vapor, $HgCl_2$ vapor, CH_3HgCl vapor, and $(CH_3)_2Hg$ have been found to exist alone or adhere to other particles.

There have been many studies on the collection of organic lead samples in the atmosphere. Fiberglass and membrane filtration are widely used to collect non-volatile inorganic lead salt, and alkyl lead compounds in the air can be filtered through these filters. Some adsorbents and extractants have been boldly used for the collection and separation of organic lead. Iodine chloride is an effective adsorbent. Dithizone chelate of dialkyl lead can be extracted by carbon tetrachloride under the action of EDTA to distinguish organic lead from inorganic lead, and inorganic lead remains in solution. Separation of organic lead with chromatographic fillers is also a common choice.

The collection of mercury-containing samples in the atmosphere includes different gas traps or adsorption methods. Different adsorbents can selectively adsorb various forms of mercury. For example, $HgCl_2$ can be adsorbed by silanized Chromosorb W, CH_3HgCl can be adsorbed by Chromosorb W treated with 0.5moL/L NaOH, elemental mercury can be adsorbed by silver-coated glass beads, and gold-coated glass beads can selectively adsorb $(CH_3)_2Hg$. For the collection of atmospheric samples with very low content, it takes several hours with a fast sampler at the flow rate of 5-50m^3/min. The disadvantage of this method is that there is the possibility of desorption or volatilization of the adsorbed samples. With a low-speed sampler at the flow rate of 0.5-1.5m^3/min, sometimes it will take several days.

Chapter 3 Pretreatment technology of environmental samples

2 Collection of indoor air pollutants samples

The main components of indoor air pollutants include formaldehyde, benzene, toluene, xylene, ethylbenzene, and other volatile toxic compounds. Indoor air sampling methods are mainly divided into direct sampling method and dynamic sampling method. When the indoor air pollution is serious or the detection method is very sensitive, direct sampling can be selected. Common tools for direct sampling include syringes, sampling bags, gas pipelines and vacuum cylinders. The extraction fibers of solid phase microextraction (SPME) developed at present can also be used for field direct sampling. In the determination of indoor air pollutants by gas chromatography, 100-500mL syringes or sampling bags are usually used. After directly extracting air samples, the closed inlet is rotated, and samples are brought back to the laboratory and directly injected into the gas chromatography inlet for analysis. In order to collect indoor air with special plastic bags, plastic bags should be selected first according to the characteristics of the samples collected. The commonly used sampling plastic bags are polyvinyl chloride bags, polyethylene bags and polytetrafluoroethylene bags. Selection of bags with metal film lining (mainly aluminum lining) is beneficial to the stability of the samples. For mixed samples containing CO, they can only be stabilized for more than ten hours in polyvinyl chloride bags. Same samples, placed in polyester bags lined with aluminium film, can be stabilized for 100 hours without loss. When using plastic bags to collect indoor air, a large volume sampler can be used. But the best choice is to use double ball. Two-ball method is suitable for collecting 100-500mL samples. Gas samples are also collected in fixed volume containers. Common containers are vacuum gas cylinders made of pressure-resistant transparent glass or stainless steel. Their volumes are 500-1000mL.

Dynamic sampling is often used when the concentration of pollutants is low or samples are collected continuously. Dynamic sampling method uses mechanical pumps to force indoor air to collect samples through the medium, so that the measured substances are adsorbed and condensed in the medium to achieve the purpose of enrichment sampling. Absorbing media can be either liquid, filled column or various membrane materials. Absorption method of solution use a gas absorption tube with absorbent liquid in it. The back end of the tube is connected with an air extraction device and the air sample is pumped through the absorbent tube at a certain flow rate. When the air passes through the absorbent solution, due to the dissolution or chemical reaction, the components tested are left in the absorbent solution. After sampling, the absorbent liquid is poured out. Determine the content of the constituents. Filled column adsorption method is widely used. Activated carbon and silica gel are commonly used as column adsorption materials. Effective adsorbents can be selected according to the different tested objects. For mercury-containing samples, Tenax of 60-80 mesh can be selected. The filled column sampling method is the most widely used method at present. Compared with the solution absorption method, this method has four characteristics: ①It can be sampled continuously for a long time, and can effectively reflect the dynamic process of pollution changes by measuring the concentration changes in different periods and sections; ②By choosing the appropriate filler, the collection efficiency of gas, steam and aerosol is better; ③The pollutants are relatively stable on the filled column and easy to be stored or carried; ④ It is suitable for various occasions, including field sampling. The pumps commonly used in air samplers include membrane pumps, electromagnetic pumps and scraper pumps. Low temperature condensation sampling method is often used to collect volatile gases. Commonly used refrigerants are ice-brine ($-10℃$), dry ice-ethanol ($-72℃$), liquid oxygen ($-183℃$), liquid nitrogen ($-196℃$) and semiconductor refrigerators (up to $-50℃$). In addition, there are passive sampling methods.

Section 2 Collection and preservation of water samples

1 Collection of water samples

Water samples are the most frequently collected in the determination of environmental samples. Environmental water samples can be divided into natural water (rain and snow water, river water, lake water, sea water, etc.), industrial wastewater, and domestic sewage. Water in nature contains complex components, including organic colloids, bacteria and algae. Inorganic solids include metal oxides, hydroxides, carbonates and clays. The content of trace elements or organic pollutants is often very low. The water samples collected must be representative.

1.1 Collection site

There are many toxic compounds in industrial sewage, while organic matter, nutrients and other components in domestic sewage are in the majority. Sampling should take into account all the influencing factors and the possible changes of these factors as far as possible. It mainly includes the following factors: ①Determination content, i.e., the category of determinated compounds. Before collection, the use of the sample should be clearly understood. If the long-term variation of a pollutant in a river is determined, the site that can be repeatedly selected within a fixed interval must be selected. ②The approximate concentration range of the sample. ③Matrix types and homogeneity. ④Special requirements of the analytical methods used. Physical processes affecting the properties of water samples include gas escape, precipitation, suspended matter damage, sediment and suspended matter disturbance, redistribution of analytes and contamination, etc. The main chemical processes affecting the properties of water samples include chemical degradation and light degradation, etc. Avoid contamination of sampling equipment, ship deck or sewage when sampling.

The content of trace elements and organic pollutants in nature is related to the depth of water samples, salinity and emission sources.

All kinds of water samples collected must be representative and can reflect the water quality characteristics. Before the detection sections of estuaries and harbours are laid out, river flow, pollutant types, point or non-point pollutant sources, discharge types of pollutants from direct sewage outlets, and other factors affecting the uniformity of water quality should be checked. The layout of monitoring section should be representative, which can truly and comprehensively reflect the spatial distribution and variation of water quality and pollutants. For the collection of water samples discharged by pipelines or canals, the uniformity of pollutant distribution must be determined first by experiments. Sampling from edge, surface or ground should be avoided, because the samples from these parts are usually not representative. Samples for analysis of the chemical composition of natural water in rivers are usually sampled 0.2-0.5m below the surface of the water in the section of hydrological station. Sampling points should be added when the section is open. Sampling points on shore must be located where the water flow is clear. Sampling sections in the estuary area are generally vertical to the direction of runoff diffusion. The sampling section (station) of the harbour is laid out according to the topography, tide, channel and detection objects. In the complex area of tidal current, the sampling section can be set vertically with the shoreline. Sampling stations in open coastal areas are arranged in a grid of longitudinal and transverse sections. If necessary, stratified sampling can be made at different depths according to different physical and hydrological characteristics and sampling requirements. Generally, it can be divided into surface layer, 10m layer and bottom layer.

1.2 Sampling requirements

In general, special samplers should be used to collect representative water samples from specified water depths.

① The collection of surface water samples must consider inserting polyethylene bottles below the surface of the water, avoiding the surface of the water and putting on polyethylene gloves. The surface water samples can be collected with polyethylene bucket. When determining metal elements or organic pollutants in seawater, more attention must be paid to the cleaning of sampling equipments. When collecting water samples with ships, we must consider the contamination from the hull itself, whether it is a big ship or a small boat.

② For deep water sampling, most of the utensils currently used are processed from polyethylene, polypropylene, polytetrafluoroethylene and organic glass (methyl methacrylate). Avoid the use of rubber rope, wire rope and other materials containing rubber or metal, and avoid rust or grease contamination.

③ For natural water samples, the method of timing collection is mostly used. In order to reflect the overall quality of water, repeated sampling must be carried out at different locations and time. In addition, in the presence of multiple emission sources, samples taken from different cross sections or depths will vary greatly. Automatic acquisition devices are mainly used for high sampling density and long-term continuous sampling requirements. The conventional parameters of continuous measurement mainly include pH, conductivity, salinity, hardness, turbidity, viscosity and so on.

④ When collecting rainwater and snow samples, if it is sediment, wet and dry sediments can be collected simultaneously by mass samplers. Wet samples can only be collected when it rains or snows. For the collection of alpine show and polar snow, clean polyethylene containers must be used. The operators wear clean gloves, and the samples should be taken against the wind. A slope about 30cm deep is scraped out with a plastic shovel. Snow samples 15-30cm away from the ground are collected horizontally with a polyethylene bottle of about 1000mL. The samples are immediately sealed and refrigerated until the samples are analyzed.

1.3 Sampling frequency

The determination principle of sampling time and frequency is to satisfy the information needed to reflect environmental information with minimum workload, to reflect the changing characteristics of environmental factors, and to consider the continuity of sampling time, technical feasibility and possibility as far as possible. The sampling frequency must be large enough to reflect the seasonal variation of water samples. Twice a week or twice a month is usually used. Sampling can also vary with the emission time of sources.

2 Pretreatment and preservation of water samples

Unless the sampled water is analyzed immediately, proper pretreatment is required before storage. Pretreatment mainly according to the different requirements of the measured water samples. Usually for trace elements or organic analysis, the particulate matter in the water sample must be removed by filtration or centrifugation (if the pollutant composition in the particulate matter is determined, this part of the sample needs to be collected), then a protective agent is added, and the water sample is stored in a uncontaminated container and stored at a suitable temperature to avoid the loss, degradation or morphological change of active ingredients.

In unfiltered samples, due to the interaction between particulate matter and debris dissolved in samples, it is possible to change the chemical speciation distribution of heavy metals in samples. Researchers found that the adsorption-desorption equilibrium time of heavy metals in the mixture of sediment and water is very short, usually less than 72hours. The maximum adsorption occurs at pH=7.5. After sampling, any change of solution equilibrium will result in that the adsorption sites provided by particulates will provide pathways for metal morphology migration. Under certain conditions, desorption of adsorbed metals is possible. In addition, some studies have shown that the unfiltered seawater samples are stored in polyethylene containers, and the dissolved heavy metal components such as Pb, Cu, Cd, Bi are not lost.

High concentration of bacterial accompanied by the presence of sediment can also lead to the loss of water-soluble metal morphology. The growth of bacteria and algae, including photosynthesis and oxidation, will change the content of CO_2 in water samples and lead to the change of pH value. The change of pH value often leads to precipitation, change of chelation or adsorption behavior, and the redox action of metal ions in solution.

Using untreated membranes to filter mercury-containing seawater samples may result in 10%-30% loss. However, the loss of mercury can be reduced to less than 7% by using treated fiberglass to filtrate.

Because of the unmeasurable nature of bacterial growth and reproduction in stored samples, the earlier the filtration is, the better. If the time is delayed to a few hours, it is better to preserve the sample at about 4℃ in order to inhibit the growth of bacteria.

The microporous membrane of 0.45μm can be used to distinguish solute from particulate matter conveniently. The filtrate through the membrane may also contain colloidal particles of microorganism and bacteria of 0.001-0.1μm and components soluble in water less than 0.001μm. The 0.45μm filter can remove all phytoplankton and most bacteria. Continuous filtration may sometimes result in blockage of the filter membrane, which usually requires replacement of the new membrane or pressure filtration.

When using the filters, we should pay attention to the part of the instrument that comes into contact with the solution, such as borosilicate glass, ordinary glass, polytetrafluoroethylene, etc. At the same time, we should also consider the type of filter, such as pinhole or pressure. The use of rubber stoppers in glass filters is liable to contamination. Vacuum filtration system using borosilicate glass is generally chosen. Before filtration, the filter material is washed with dilute acid and can usually be soaked overnight in 1-3mol/L hydrochloric acid.

Cadmium and lead are easily adsorbed on the surface of untreated filter membranes, but no change in the concentration of these elements is found when membranes are used to filter river water. Usually, the membrane is washed with 20mL 2mol/L HNO_3 before use, and then washed with 50-100mL distilled water. The received beaker or flask must be washed with distilled water, and the initial 10-20mL filtrate should be removed. For the filtration of deep sea samples, it is better to immerse the membrane in dilute nitric acid first.

Pressure filtration or vacuum suction filtration are two commonly used methods. Pressure filtration is fast and suitable for the filtration of river water samples containing a large amount of sediment. If 47mm diameter, 0.45μm membrane is used to filter water samples, the speed is about 100mL/h. Ultrafiltration membrane is usually used for pressure filtration.

Centrifugation is also an effective method for hard-to-filter samples, but the process of centrifugation can easily cause contamination. The efficiency of centrifugal separation depends on the speed, time and density of particle.

After water sample is collected, it should be analyzed as soon as possible. If it is placed too long, some components in water will change. The allowable storage time of water sample for physical and chemical analysis is as follows: clean water, 72hours; slightly polluted water, 48hours; polluted water, 12hours. If water samples cannot be transported in time or analyzed as soon as possible, appropriate preservation methods should be adopted according to the requirements of different testing items. Preservation technology of common water samples are as show in table 3-2-1.

Table 3-2-1 Preservation technology of common water samples

Project to be tested		Container class	Preservation method	Storable time	Advice
Physical and chemical analysis	pH	P/G	—	—	Field direct test
	Acidity/Alkalinity	P/G	Refrigerate in 2-5℃ dark place	24h	Fill the container with water sample
	Smell	G	—	12h	—
	Electrical conductivity	P/G	Refrigerate at 2-5℃	24h	—
	Chroma	P/G	Refrigerate in 2-5℃ dark place	24h	—
	Suspended solids	P/G	—	24h	Single constant volume sampling
	Turbidity	P/G	—	—	Field direct test
	Ozone	G	—	—	—
	Residual chlorine	P/G	Fix with NaOH	6h	Best on-site analysis
	Carbon dioxide	P/G	—	24h	Fill the container with water sample
	Dissolved oxygen	—	Fix and store in the dark on site	—	Iodometry with 1mL 1mol/L manganese sulfate solution and 2mL 1mol/L alkaline potassium iodide
	Grease/Oil	G	Acidify with HCl to pH≤2	1 week	—
	Hydrocarbons/Petroleum and derivatives	G	Acidification with HCl or H_2SO_4, pH=1-2	1 month	—
	Potassium permanganate index	G	Refrigerate in 1-5℃ dark place	24h	—
		P	Freeze at −5℃	1 month	—
	COD	G	Acidify with H_2SO_4 to pH≤2	24h	—
		P	Freeze at −20℃	1 month	—
	Arsenic	P/G	Add 2mL concentrated HCl to 1L water sample	2 weeks	—
	Biochemical oxygen demand	G	Refrigerate in 2-5℃ dark place	As soon as possible	—
	Kjeldahl nitrogen, ammonia nitrogen	P/G	Acidify with H_2SO_4 to pH<2, and refrigerate at 2-5℃	As soon as possible	—
	Nitrate nitrogen	P/G	Acidify to pH<2 and refrigerate at 2-5℃	As soon as possible	—
	Nitrite nitrogen	P/G	Refrigerate at 2-5℃	As soon as possible	—
	Organochlorine pesticide	G	Refrigerate at 2-5℃	1 week	—

Chapter 3 Pretreatment technology of environmental samples

Continued

Project to be tested		Container class	Preservation method	Storable time	Advice
Physical and chemical analysis	Organophosphorus pesticide	G	Refrigerate at 2-5℃	24h	—
	Phenol	P	Inhibition of biochemical action with copper sulfate, acidification with phosphoric acid, or adjustment with NaOH to pH＞12	24h	—
	Chlorophyll a	BG	Refrigerate at 1-5℃	24h	—
	Mercury	P/G	—	2 weeks	—
	Cadmium, lead, copper, aluminum, manganese, zinc, nickel, total iron, total chromium	P/BG	Acidify with nitric acid to pH＜2	1 month	—
	Hexavalent chromium	P/G	Adjust with NaOH to pH＝7-9	—	—
	Calcium, magnesium, total hardness	P/BG	Acidify the filtrate after filtration to pH＜2	A few months	Do not use H_2SO_4 when acidizing
	Fluoride	P	Neutral samples	A few months	—
	Chloride	P/G	—	A few months	—
	Total phosphorus	BG	Acidify with H_2SO_4 to pH＜2	—	—
	Selenium	G/PG	Adjust with NaOH to pH＞11	A few months	—
	Sulfate	P/G	Refrigerate at 2-5℃	1 week	—
Microbiology and Biological analysis	Total bacterial/Coliform/ Salmonella, etc	Sterilizing container G	Refrigerate at 2-5℃	As soon as possible	—
	Identification and counting: benthic invertebrates	P/G	Add 70% (volume fraction) ethanol or 40% (volume fraction) neutral methanol	1 year	—
	Phytoplankton and zooplankton	G	Add 40% (volume fraction) formaldehyde to form a 4% (volume fraction) formalin solution, or add 100copies of Lugol solution to a volume sample.	1 year	Lugol solution: 150g KI, 100g I_2, and 10mL ethanol as aqueous solution

Note: P—Polyethylene; G—Glass; BG—Borosilicate glass.

Section 3 Collection and preservation of soil samples

1 Collection of soil samples

1.1 Collection of polluted soil samples

(1) Sampling distribution point

A certain number of plots representing the investigated area are selected as sampling units (1300-2000m^2). A certain number of sampling points are set up in each sampling unit. At the same time, the control sampling unit is selected to set up sampling points.

In order to reduce the influence of heterogeneity of soil spatial distribution, multi-point sampling should be carried out in different directions in a sampling unit. And evenly mixed them into representative soil samples.

(2) Sampling depth

Generally, soil pollution is monitored by taking 0-15cm or 0-20cm topsoil and sampling with a shovel.

(3) Sampling quantity

Soil samples obtained by the above methods are usually mixed with the average amount of multiple points, and the amount of soil taken is usually large, but only 1-2kg is needed. Therefore, the obtained mixed samples must be discarded repeatedly according to the quartering method. Finally, the amount of soil sample needed is left and packed in plastic bags or cloth bags.

1.2 Sample collection of soil background value

Soil profiles need to be excavated for sampling at each sampling point. China environmental background value research cooperative group recommended that the profile specifications are generally 1.5m long, 0.8m wide and 1.0m deep, and each profile collects soil samples from three layers A, B and C. Transition layers (AB, BC) are generally not sampled. When the groundwater level is high, it is excavated until the groundwater is exposed. The actual sampling depth is recorded, such as 0-20cm, 50-65cm and 80-100cm. Sampling from the bottom to the top at the typical center of each level should not confuse the level and mix samples.

In areas with thin soil layers in mountainous areas, only A and C layers are taken when B layers are incomplete.

Samples near surface (0-20cm), middle (50cm) and bottom (100cm) are collected from incomplete soil in arid areas.

Determination of sampling points: The number of sampling points is related to the size of the study area and the precision of the research task. In order to make the distribution of soil background values more reasonable, sampling points are based on statistical principles, that is, at the selected confidence level, they are related to the standard deviation of measured values and the required accuracy. The number of sampling points per sampling unit can be estimated as follows:

$$n = t^2 s^2 / d^2 \qquad (3\text{-}3\text{-}1)$$

Where, n——the minimum number of sampling points involved in each sampling unit;

t——the confidence factor (when the confidence level is 95%, t is 1.96);

s——the relative standard deviation of samples;

d——the allowable deviation (if the sampling accuracy is not less than 80%, d is 0.2).

2　Pretreatment and preservation of soil samples

1.1　Air drying of soil samples

After the soil samples collected from the field are transported to the laboratory, in order to avoid mildew and deterioration caused by microorganism, all the samples should be immediately poured on the plastic film or into ceramic dishes for air drying. When it reaches the semi-dry state, crush the soil, remove the stones, residual roots and other debris and lay it into a thin layer, often turn over, make it slowly air-dry in the shade, and avoid direct sunshine exposure. The air-drying place of the sample should prevent the pollution of acid, alkali, and other gas and dust.

1.2　Grinding and screening

Usually, the fineness of the sample is determined by the measured components and the weighing of the sample.

(1) Physical analysis

The air-dried samples are collected with 100-200g and crushed with a log stick on the board. After repeated treatment, all the soil samples pass through a 2mm pore sieve. The soil samples are mixed and stored in a wide-mouth bottle for soil particle analysis and physical properties determination.

(2) Chemical analysis

For the analysis of organic matter and total nitrogen projects, a part of soil samples which have passed 2mm sieve should be taken and further refined with an agate grinding bowl, so that all of them pass through sieve No. 60 (0.25mm). When measuring heavy metals such as Cu, Cd and Ni by atomic absorption spectrophotometry (AAS), all soil samples must be ground and sifted through sieve No. 100 (nylon sieve), and then mixed, bottled, labeled, numbered and stored.

1.3　Preservation of soil samples

General soil samples should be kept for six months to one year for checking if necessary. Standard soil samples or control soil samples used for quality control in environmental monitoring need to be well preserved for a long time. Samples should be stored to avoid the effects of sunlight, humidity, high temperature and acid-base gases.

Store air-dried soil samples, sediments or standard soil samples in clean glass or polyethylene containers. It is feasible to preserve for 30months under normal temperature, cool, dry, sunproof and sealed (paraffin-coated) conditions.

Section 4 Collection and preservation of biological samples

1 Collection of biological samples

Biological samples involve complex matrices, both solid and liquid, including all aquatic or terrestrial animals and plants. Morphological analysis is sometimes directed at the whole organism, sometimes part of its organs or components, and sometimes only the determination of excreta. In environmental analysis, biological samples mainly include fish, fruit shells, algae, herbs, fruits, vegetables, leaves and other animal and plant samples. In occupational health research, we mainly study human tissues, hair, sweat, blood, urine and feces.

The key of collecting biological samples is to prevent contamination. For the speciation analysis of metal elements, metal scrapers, dissectors, scissors, tweezers or needles should be avoided. Borosilicate glassware for laboratory use, polyethylene, polypropylene or polytetrafluoroethylene, and quartz appliances can replace the above mentioned metal products. When handling samples, do not directly contact with hands, but should wear plastic gloves. Powder such as talc powder should be washed clean, because these powders often bring zinc and other metals. In practice, metal blades and needles for biopsy are inevitably used for sample processing. Although these tools may cause some metal contamination, they are much more convenient and effective than quartz and glass knives. The results show that it is possible to bring 3ng/g Mn, 15 ng/g Cr and 60ng/g Ni into the biopsies by using stainless steel tools. If we study the concentration and morphology of the above metal elements in the biological body, it is obviously inappropriate to use stainless steel tools.

In the study of trace elements in human serum, sometimes the content is quite different. This is due to the pollution of sampling to a large extent, besides the differences between analytical methods and individuals. Some studies have shown that the content of iron, manganese, nickel, cobalt, molybdenum and chromium in the first 20mL blood samples collected by stainless steel needles has increased significantly, while the content of copper and zinc has no change. For this reason, polytetrafluoroethylene and polypropylene catheters are preferable alternatives.

Dust is the main source of zinc, aluminum and other metal pollution in biological samples. Dust should be avoided in the collection of body fluids and other samples. Urine samples are also highly susceptible to dust, so more attention must be paid to sampling and receiving them in capped polyethylene bottles washed with acid.

2 Pretreatment and preservation of biological samples

For many biological samples, initial pretreatment is required. This pretreatment should be done immediately after sampling. For example, in the analysis of trace elements in shell samples, it is necessary to clean the sediments in the outer layer of the shell. Then open the shell and collect the whole flesh or individual parts. Similarly, aquatic plants such as algae need to be carefully cleaned to remove sediments, parasitic plants or other similar surface contamination. We must pay attention to the representativeness of samples. Trace elements tend to have higher concentration in some special parts, such as plant roots and leaves, and the size of the plant is also related to the concentration. If a single sample is analyzed, attention should be paid to these characteristics when sampling. The above samples should be refrigerated if they are not analyzed immediately after collection. Polyethylene gloves should be worn when handling the above samples. Samples should be stored in plastic bags or containers. For hair samples, in order to remove foreign elements, some cleaning methods can be used for trace element analysis. In 0.1% Triton-X100, the hair samples can be treated with ultrasonic oscillator, filtered and washed with methanol, and then dried with a hair dryer in the air.

For the treatment of blood samples, pretreatment should be carried out according to different requirements of whole blood, plasma and serum. Whole blood can be stored at 4℃ in a short period of time. The effect of freezing is better, and the cold storage often leads to precipitation. The solid precipitated during thawing can be dissolved by adding acid.

Sample homogenization is often the second step in processing a large number of samples. If a mechanical homogenizer is used, the problem of blade contamination needs to be considered. A small amount of samples can also be dissolved directly in concentrated hydrochloric acid or strong alkali. Quaternary ammonium hydroxides are also often used to dissolve small amounts of tissue samples. If soaked in 20% tetramethylammonium hydroxide for 2hours, alkyl lead compounds can be effectively separated from protein and fat. This extraction method is suitable for the treatment of fish, algae and other marine organisms. The extraction can be accelerated by heating in 60℃ water bath.

High-pressure tanks are also often used for sample processing. This method can effectively prevent contamination during sample processing to protect trace elements. Before dissolving samples under high pressure, the samples are usually freeze-dried or air-dried, and then treated under high pressure. The temperature control of the high-pressure tanks depends on the requirements of different samples. Generally speaking, the satisfactory results can be obtained after 6h of treatment at 160-180℃.

Freeze-drying is a good method to remove moisture from solid biological samples, which can avoid the loss of trace elements or sample contamination. A small number of samples are suitable for this method. Controlling the temperature in the fume hood to heat samples continuously at about 100℃ is the most convenient method, but volatile components, such as elemental mercury and some other organic metal compounds, are easily lost during this period.

　　Muffle furnace can decompose most organic compounds at 500℃. It is also suitable for the treatment of samples containing Hg, As, Sn, Se, Pb, Ni, Cr and others.

　　Freeze-dried or digested samples are easier to preserve and need to be mixed uniformly again before extraction or dissolution.

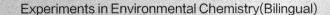
Experiments in Environmental Chemistry(Bilingual)

· Chapter 4 ·
Environmental chemistry experiments

Section 1 Atmospheric environmental chemistry experiments

Experiment 1 Pollution analysis of PAHs in urban atmospheric aerosol

Polycyclic aromatic hydrocarbons (PAHs) which are ubiquitous in the urban atmosphere, are listed as persistent organic pollutants that need to be controlled and treated due to their high toxicity, persistence, accumulation and high mobility.

PAHs are compounds in which two or more benzene rings are connected in a fused ring. According to their molecular structure, they can be divided into two categories, namely biphenyls and fused-ring compounds. Polycyclic aromatic hydrocarbons and heterocyclic compounds are usually synthesized by high temperature combustion. The PAHs in atmospheric aerosols mainly come from incomplete combustion of carbonaceous substances. The pollution source of vehicle exhaust is the main contributor of PAHs in PM10, followed by coal burning. The PAHs in the dust mainly come from the deposition of coal dust. At present, most of the studies on PAHs in air in our country focus on PAHs in particulate matter, but relatively fewer on PAHs in atmospheric aerosol. In this experiment, the medium flow sampling method is used to measure PAHs in atmospheric aerosol, and the existence state of PAHs in atmospheric environment, namely the distribution of PAHs in atmospheric aerosol, is analyzed.

1 Experimental purpose

(1) Master the collection, extraction and analysis methods of PAHs in the air.

(2) Master the determination principle and application method of HPLC.

(3) Analyze and evaluate the pollution status and existence form of PAHs in the air.

2 Experimental principle

The main forms of PAHs in air include gas state, particle state (adsorbed on particulate matter), and under certain conditions, they can be converted into each other. Factors affecting PAHs content include physical and chemical morphology, temperature, and other co-existing pollutants, such as dust and ozone.

The main steps are as follow:

(1) PAHs in atmospheric aerosol are collected by automatic atmospheric monitoring sub-station;

(2) As an extractant, dichloromethane was used to extract PAHs from samples by ultrasound.

(3) The peak height or peak area of PAHs is determined by HPLC. PAHs are quantified by the external standard method.

(4) Analyze and evaluate the pollution level and morphological distribution of PAHs.

3 Experimental instruments and reagents

3.1 Instruments

High performance liquid chromatograph, with LC7060 type photodiode array (PDA) detector, ultrasonic cleaners, air automatic monitoring sub-station, RAM-1020 β-ray dust meter, K-D concentrator, $0.45 \mu m$ membrane, glass fiber filter membrane belt (30mm × 20m).

3.2 Reagents

PAHs standard mixed solution (100mg/L) whose composition is acenaphthene (Ace), fluorene (Fl), phenanthrene (Ph), anthracene (An), fluoranthene (Fa), pyrene (Py), benzo[a]anthracene (BaA), chrysene (Chry), benzo[b]fluoranthene (BbF), benzo[k]fluoranthene (BkF), benzo[a]pyrene (BaP), dibenz[a,h]anthracene (DahA), benzo[g,h,i]perylene (BghiP) and indeno[1,2,3-c,d]pyrene (IcdP), which are $200 \mu g/mL$. (milimbo); Mili-Q pure water, ultrasonic degassing for later use; methanol (chromatographic purity); nitric acid (excellent purity); dichloromethane (chromatographic purity).

4 Experimental procedure

4.1 Sampling point selection

Select sampling points according to actual conditions in the following 5 different functional areas:

(1) Industrial area, choose 5-15 sampling points near the chemical plant;

(2) Residential area, 10-15 sampling points should be selected near relatively representative residential area;

(3) Business district, 5-15 sampling points should be selected near representative commercial streets;

(4) Traffic dense area, choose 5-10 sampling points near the main roads of the city;

(5) Cultural and educational district, choose representative schools, museums and other places related nearby with 5-10 sampling points.

4.2 Pretreatment of filter membrane belt

Before use, the glass fiber filter membrane belt is soaked in dilute nitric acid solution for one week, then soaked in Mili-Q water for one week, then dried naturally and dried in a dryer until the filter membrane is constant weight and ready for use. The blank value of filter membrane treated by this method is close to 0.

4.3 Sample collection

The atmospheric aerosol samples are collected by RAM-1020 ray dust meter equipment on each automatic atmospheric sub-station. Continuous aerosol samples are collected with a 20mm wide and 20m long glass fiber filter membrane belt for 48h, with the sampling head 9-18m above the ground. The number of atmospheric aerosol samples collected at the sampling points near each functional area varies from 15 to 40 according to the analysis requirements.

4.4 Sample pretreatment

Cut the collected filter membrane belt with spots into pieces and put them into the 25mL grinding conical flask, add a certain amount of dichloromethane, use ultrasonic extraction 3 times and each time is about 5min. The extract is filtered through 0.45μm filter membrane into K-D concentrator. Try to concentrate liquid product to about 0.5mL on the 65-70℃ water bath, then dry with pure nitrogen. The residue is dissolved with methanol and the volume is fixed to 0.50mL to be measured. For HPLC analysis, filter solution before injection with a 0.45μm needle filter.

4.5 Chromatographic operation procedure

Analysis of PAHs by high performance liquid chromatography uses unfixed fluorescence excitation wavelength and fluorescence emission wavelength of high performance liquid chromatography-fluorescence detection method to determine the PAHs in atmospheric aerosol. This method with the advantages of high sensitivity and less interference is simple and fast for actual samples. The analytical wavelengths of PAHs are shown in table 4-1-1.

Table 4-1-1 Analysis wavelength of PAHs

Time/s	Excited wavelength, λ/cm	Emission wavelength, λ/cm	Determination components
0	291	356	Acenaphthene(Ace), fluorene(Fl)
210	250	400	Phenanthrene(Ph), anthracene(An)
305	289	462	Fluoranthene(Fa)

Continued

Time/s	Excited wavelength, λ/cm	Emission wavelength, λ/cm	Determination components
355	320	380	Pyrene(Py)
540	266	403	Benzo[a]anthracene(BaA), chrysene(Chry), benzo[b]fluoranthene(BbF), benzo[f]fluoranthene(BkF), benzo[a]pyrene(BaP)
900	294	430	Dibenzo[a,h]anthracene (DahA), benzo[g,h,i]perylene (BghiP)
1530	294	482	Indeno[1,2,3-c,d]pyrene (IcdP)

The column was PECHROMSEP, 250mm×4.6mm. The flow rate of mobile phase is 1.0mL/min. The flow of mobile phase is approximately 2L. The mobile phase is methanol-water and gradient elution is performed. The optimal volume ratio of methanol-water is methanol (15%) : water (100%) = 18 : 82, which was maintained at this ratio for 7min. After linear gradient elution for 5min, the ratio is changed to: methanol (15%) : water (100%) = 5 : 95, which is maintained for 17min.

Plot of standard curve: The peak height is the ordinate and the concentration of PAHs is the abscissa. Plot the standard curve of each kind of PAHs. The range of concentration should be determined according to the sensitivity of HPLC and the concentration of the sample.

5 Data processing

Calculation the concentration of PAHs in atmospheric aerosol: according to chromatogram, retention time is qualitative, peak height and peak area are quantitative, and the concentration is calculated by integral. The linear gradient of PAHs mobile phase is shown in table 4-1-2 below, which can be selected according to the actual sample concentration.

Table 4-1-2 Linear gradient of PAHs mobile phase

Time/min	Methanol(volume fraction)/%	Water(volume fraction)/%
0.0	50	50
5.5	70	30
16.0	80	20
20.0	85	15
25.0	90	10
30.0	95	5
35.0	95	5
40.0	100	0
45.0	100	0
50.0	50	50

6 Questions

(1) Analyze the experimental data and explain the main sources of PAHs in the air.

(2) Describe the main factors affecting the existence form of PAHs in the air.

(3) Several methods for the determination of PAHs and their differences are illustrated.

Experiment 2 Pollution analysis of benzene in indoor air

1 Experimental purpose

(1) Learn the basic operation of gas chromatography.

(2) Learn how to establish a chromatographic method for the determination of benzene in indoor air.

(3) To understand the influence of indoor ventilation conditions on benzene content in indoor air.

2 Experimental principle

Benzene is a kind of relatively common pollutant. Benzene in indoor air usually comes from solvent volatilization and organic matter combustion of indoor products. "Standards for indoor air quality" stipulates that the limit of benzene in indoor air is $0.03mg/m^3$.

Benzene is usually colorless, aromatic and volatile. It is slightly soluble in water and easily soluble in organic solvents such as ethanol, ether, chloroform and carbon disulfide. The concentration of benzene in indoor air can be determined by activated carbon adsorption sampling or low temperature condensation sampling, and then determined by gas chromatography. Determination methods and characteristics of benzene series are shown in table 4-1-3. DNP-Bentane column (CS_2 desorption) method is used in this study.

Table 4-1-3 Determination methods and characteristics of benzene series in indoor air

Determination method	Principle	Range of measurement	Characteristics
DNP+Bentane (CS_2 desorption)	After enrichment of benzene, toluene, ethylbenzene and xylene in air with activated carbon adsorption sampling tube, CS_2 is added, separated by DNP + Bentane chromatographic column and determined by flame ionization detector. The retention time is qualitative and peak height (peak area) is quantitative by external standard method	When the peak area is 100L, the minimum detected concentration is $0.006mg/m^3$ benzene, $0.004mg/m^3$ toluene, $0.010mg/m^3$ xylene and ethylbenzene	It can separate and determine acetone, styrene, ethyl acetate, butyl acetate and amyl acetate in air at the same time. The measurement is extensive

Determination method	Principle	Range of measurement	Characteristics
PEG-6000 column (CS_2 desorption injection) method	Benzene, toluene and xylene in the air are collected with activated carbon tube, and samples are desorbed with CS_2. After being separated by PEG-6000 column, hydrogen flame ionization detector is used to detect the samples. The retention time is qualitative and the peak height is quantitative	The detection limits of benzene, toluene and xylene are respectively 0.5×10^{-3}, 1×10^{-3} and 2×10^{-3} μg (inject 1μL liquid sample)	Can only detect benzene, toluene, xylene, styrene
PEG-6000 column (thermal desorption injection) method	Benzene, toluene and xylene in the air are collected with activated carbon tube, and samples are injected after thermal desorption. After being separated by PEG-6000 column, hydrogen flame ionization monitor is used to detect the samples. The retention time is qualitative and the peak height is quantitative	The detection limits of benzene, toluene and xylene are respectively 0.5×10^{-3}, 1×10^{-3} and 2×10^{-3} μg (inject 1μL liquid sample)	Desorption is convenient and has high frequency
Dinonyl phthalate-organic bentonite column	Benzene, toluene, and xylene gas samples are concentrated and enriched at $-78°C$, separated by dinonyl phthalate-organic bentonite chromatography column, and determined by hydrogen flame ionization detector	Detection limits: $0.4 mg/m^3$ benzene, $1.0 mg/m^3$ xylene (1mL gas sample)	The sample is unstable and needs to be analyzed as soon as possible

3 Experimental instruments and reagents

3.1 Instruments

(1) Gas chromatograph with hydrogen flame ionization detector.

(2) Chromatographic column: 2m × 3mm stainless steel column coated with 2.5% Bentane Chromosorb W HPDMCS (80-100mesh).

(3) Air sampler: Flow rate is 0.2-1L/min.

(4) Microsyringe: 1 syringe, 10μL.

(5) Volumetric flask: 10 for 5mL and 100mL respectively.

(6) Straw: Several, 1-20mL.

(7) Sampling tubes: 15 glass tubes with the length of 15cm and the inner diameter of 8mm are installed with 20-50mesh granular activated carbon 0.6g (activated carbon was burned for 3h at 350°C in the muffle furnace in advance for cooling backup). Or 15 glass tubes are blown with nitrogen for 10min at 300-350°C and then divided into two sections A and B, separated by glass wool in the middle and sealed at both ends before use.

3.2 Reagents

(1) Benzene (chromatographic pure reagent).

(2) Carbon disulfide (CS_2): It must be purified before use (see "matters needing attention" for purification method) and tested by chromatography. Inject $5\mu L$. There is no peak between benzene and toluene shows it can be used.

(3) Benzene standard reserve solution: Add 90mL purified CS_2 solution into a 100mL volumetric flask. Then, add $10.0\mu L$ benzene solution and dilute the solution to the mark with CS_2 solution to obtain benzene standard solution with the concentration of $88\mu g/mL$. Next, add 80mL CS_2 solution into another 100mL volumetric flask. Then, add 10.0mL benzene standard solution with the concentration of $88\mu g/mL$ and dilute the solution to the mark with CS_2 solution to obtain benzene standard reverse solution with the concentration of $8.8\mu g/mL$. This reverse solution can be stored at 4℃ for 1 month.

4 Expermental procedure

4.1 Sampling

Connect the sampling tube mouth with the air sampling pump inlet with latex tube and place it vertically. At the flow rate of 0.5L/min, sample 40L gas in the window (indoor side). After sampling, the two ends of the sampling tube are sealed with latex tube and samples are measured within 10 days. Record the temperature and atmospheric pressure at the sampling point.

Parallel sampling: Take another two tubes for parallel sampling in accordance with above methods, Measure and calculate the average value of three measurements.

Sample at other locations in the room: Take another sample tube, place it in the doorway (indoor side), furniture, bedroom and living room, set sampling points, and sample and measure them with the same method mentioned above.

4.2 Determination

Chromatographic conditions: Column temperature is 64℃. Gasification chamber temperature is 150℃. Test room temperature is 150℃. Carrier gas (nitrogen) flow is 50mL/min. Fuel gas (hydrogen) flow is 46mL/min. Oxidant gas (air) flow is 320mL/min.

Standard curve drawing: 0.0, 5.0, 10.0, 5.0, 20.0 and 25.0mL of benzene reserve solution are respectively put into 100mL volumetric flasks, diluted with CS_2 to the mark line, and shaken well. In addition, six 5mL volumetric flasks are taken and added with 0.25g granular activated carbon and 2.00mL benzene standard solution of No. 0-5 respectively. After the samples are oscillated for 2min and placed for 20min, each sample is injected $5.0\mu L$ under the chromatographic conditions mentioned above. The retention time and peak height or peak area of the standard sample are determined according to the operating requirements of the gas chromatograph. Plot the standard curve of the relation between peak height or peak area and mass concentration of benzene.

Sample determination: Two sections of activated carbon in the sampling tube are transferred into two 5mL volumetric bottles respectively, and purified 2.00mL CS_2 is added, oscillated for 15min, placed for 20min, and then 5.0μL desorption solution is injected into the chromatograph. Record the retention time and peak height or peak area. The retention time is qualitative and peak height or peak area is quantitative.

5 Data processing

The mass concentration of benzene in air is calculated according to the formula (4-1-1):

$$\rho = \frac{m_1 + m_2}{V_n} \qquad (4-1-1)$$

Where, ρ —— the mass concentration of benzene in air, mg/m^3;

m_1 —— the mass of benzene series in A segment activated carbon desorption solution, μg;

m_2 —— the mass of benzene series in B segment activated carbon desorption solution, μg;

V_n —— the sampling volume in standard state, L.

6 Questions

(1) What are the methods to determine benzene in air? What are their advantages and disadvantages?

(2) What measures should be taken to reduce the measurement error when determining benzene in air by gas chromatography?

(3) Please compare the mass concentration of benzene in the samples taken from different sampling sites in the room and explain the influence of different ventilation conditions on the mass concentration of benzene in the room.

7 Notes

This method is also applicable to the determination of acetone, styrene, ethyl acetate, butyl acetate and amyl acetate in air. Under the above chromatographic conditions, the retention time is shown in Table 4-1-4.

Table 4-1-4 Retention time of each component

Component	Acetone	Ethyl acetate	Benzene	Toluene	Butyl acetate	Styrene
Retenion time/s	0.65	0.76	1.00	1.89	2.53	6.94
Component	Parax-ylene	Meta-xylene	Ortho-xylene	Amyl acetate	Ethyl benzene	
Retenion time/s	3.80	4.35	5.01	5.55	3.50	

(1) When the concentration of benzene in the air is about $0.1mg/m^3$, gas samples can be collected with 100mL syringe. The gas samples are concentrated at room temperature, then heated and desorbed, and determined by gas chromatography.

(2) Activated carbon and glass wool can only be used after blank test. The test method is to take glass wool and activated carbon (about 0.1g and 0.5g, respectively) of one acti-

vated carbon adsorption sampling tube, add 2mL purified CS_2, oscillate for 2min, place for 20min, and inject 5μL, and observe whether there is interference peak in the position to be measured. Activated carbon and glass wool with no interference peak can only be used, otherwise need to be processed in advance.

(3) Commercial analytical-grade CS_2 often contains a small amount of benzene and toluene, which must be purified before use. Purification method: mix 1mL formaldehyde with 100mL concentrated sulfuric acid. Take a 500mL dividing funnel, add 250mL commercially available CS_2 and 20mL formalde-concentrated sulfuric acid extract for market sale, and oscillate and stratify. After several times of extraction until CS_2 was colorless, it is washed twice with 20% Na_2CO_3 aqueous solution, and then redistilled to extract 46-47℃ fractions.

Experiment 3　Concentration level of formaldehyde in indoor air

Formaldehyde is the gas which is colorless, has a strong pungent smell, and can be combined with protein. Inhaling high concentration of formaldehyde can cause serious irritation and edema of respiratory tract, stinging eyes, headache, and bronchial asthma. Inhalation of a small amount of formaldehyde frequently can cause chronic poisoning, mucous membrane congestion, skin irritation, headache, heart palpitations, insomnia and other symptoms, seriously can also lead to cancer and genetic diseases.

Formaldehyde is a common organic pollutant in indoor air. Formaldehyde can be released from adhesives used in construction and decoration, chemical fiber carpets, paint coatings, cosmetics, cleaners, pesticides, food and cooking processes. Formaldehyde pollution in indoor air basically comes from decorating. The cases of malignant injury because of interior decoration and the volatile things such as formaldehyde in furniture has happened frequently. People spend about 80% of their time indoors, and indoor air pollution is more harmful to human health. Therefore, it is of great significance to master the analysis method, evaluation method and pollution status of formaldehyde in indoor air.

1　Experimental purpose

(1) Understand the basic principle of spectrophotometry for determination of formaldehyde in air, and master its operation steps.

(2) Master the basic methods of detection and evaluation of formaldehyde concentration level in indoor air.

2　Experimental principle

Formaldehyde in air reacts with phenolic reagent to form azine, which is oxidized by ferric ions in acidic solution to form blue-green compound. Spectrophotometry can be used to determine the absorbance of the compound and quantify formaldehyde. The reaction equation is showed as follows:

$$\text{MBTH} \xrightarrow{\begin{array}{c}H_2C=O\\ \\ Fe^{3+}\ [O]\end{array}} \begin{array}{c}A(\text{azine})\\ \\ B\end{array}$$

$$A + B \xrightarrow{Fe^{3+}\ [O]} \text{Blue-green}$$

Use 5mL sample solution. The determination range of this method is 0.1-0.5μg. When the sampling volume is 10L, the concentration can be measured range from 0.01 to 0.15mg/m^3. The sensitivity is 2.8μg per absorbance. The lower detection limit of formaldehyde is 0.056μg. When the formaldehyde content is 0.4-1.0μg per 5mL, the standard recovery rate is 93%-101%.

When sulfur dioxide coexists, the measurement results are lower. Therefore, sulfur dioxide interference can not be ignored. The gas sample can be first passed through filter with manganese sulfate filter paper to remove sulfur dioxide.

Preparation of manganese sulfate filter paper: take 10mL manganese sulfate aqueous solution with the concentration of 100mg/mL, drop it on 250cm^2 glass fiber filter paper, dry it and cut it into pieces, then put it into a 1.5cm×150mm U-shaped glass tube. When sampling, connect this tube in front of the formaldehyde absorption tube. The manganese sulfate filter paper made by this method has the efficiency of absorbing sulfur dioxide, which is greatly affected by atmospheric humidity. When the relative humidity is greater than 88%, the speed of gas recovery is 1L/min, and the sulfur dioxide concentration is 1mL/m^3, it can eliminate more than 95% of sulfur dioxide, and the filter paper can remain effective for 50h. When the relative humidity is 15%-35%, the efficiency of sulfur dioxide absorption decreases gradually. So when the relative humidity is very low, it should be replaced with a new manganese sulfate filter paper.

3 Experimental instruments and reagents

3.1 Instruments

(1) Large bubble absorption tube: The inner diameter of the air outlet is 1mm, and the distance between the air outlet and the bottom of the tube is less than or equal to 5mm.

(2) Constant current sampler: Flow range is 0-1L/min, the flow rate is stable and

adjustable, the constant current error is less than 2%, and the flow rate of the sampling system is calibrated with a foam flowmeter before and after sampling, the error is less than 5%.

(3) Colorimetric tube with plug: 10mL.

(4) Spectrophotometer: Determination of absorbance at 630nm.

3.2 Reagents

The water used in this method is double distilled water or deionized water. All reagents are generally analytically pure.

(1) Absorbent stock solution: Weigh 0.1g phenol reagent [$C_6H_4SN(CH_3)C:NNH_2 \cdot HCl$, MBTH], dissolve it with water, transfer it to 100mL measuring cylinder with stopper, and add water to the scale. Store it in a refrigerator at 4℃ for 3days. In the experiment, 5mL absorbent stock solution is taken and 95mL water is added to form the absorbent solution. Reformulation is required before use.

(2) Standard reserve solution of formaldehyde: Take 2.8mL solution with the content of 36%~38% formaldehyde, put it into a 1L volumetric flask, and add water to dilute it to the scale. 1mL solution is equivalent to about 1mg formaldehyde. The exact concentration should be measured by iodimetry.

Calibration of formaldehyde standard reserve solution:

① 10g/L ammonium ferric sulfate solution: Weigh 1.0g ammonium ferric sulfate dodecahydrate [$NH_4Fe(SO_4)_2 \cdot 12H_2O$], dissolve it with 0.1mol/L hydrochloric acid, and dilute the solution to 100mL.

② Iodine solution [$c(1/2I_2)=0.1000$mol/L]: Weigh 40g potassium iodide, dissolve it in 25mL water, and add 12.7g iodine. When the iodine is completely dissolved, the water should be filled to 1000mL. Transfer it to a brown bottle and store it in the dark.

③ 1mol/L sodium hydroxide solution: weigh 40g sodium hydroxide, dissolve in water, and dilute the solution to 1000mL.

④ Sulfuric acid solution [$c(1/2H_2SO_4)]=0.5$mol/L: Take 28mL concentrated sulfuric acid, slowly add it into water, cool, dilute the solution to 1000mL.

⑤ Sodium thiosulfate standard solution, which is commercially available standard solution.

⑥ 5g/L starch solution: Mix 0.5g soluble starch with a small amount of water to make a paste, then add 100mL of boiling water, and boil for 2-3minutes until the solution is transparent. After cooling, 0.1g salicylic acid or 0.4g zinc chloride is added for preservation.

Take 20.00mL formaldehyde standard reserve solution to be calibrated accurately and place it in a 250mL iodine measuring bottle. Add 20.00mL iodine solution and 15mL 1mol/L sodium hydroxide solution, place for 15min, add 20mL sulfuric acid solution, then place for 15min, titrate it with sodium thiosulfate solution until solution appears pale yellow, add 1mL 5g/L starch solution, continue the titration until blue fades, and record the volume of sodium thiosulfate solution. The volume of the standard solution of sodium thiosulfate used for blank titration is recorded. The concentration of formaldehyde solution is calculated by the equation (4-1-2):

$$\rho_{formaldehyde} = \frac{(V_1 - V_2) \times c \times 15}{20} \quad (4-1-2)$$

Where, $\rho_{formaldehyde}$ ——the concentration of formaldehyde standard reserve solution;

V_1 —— volume of sodium thiosulfate solution consumed by reagent blank, mL;

V_2 ——volume of sodium thiosulfate solution consumed by formaldehyde standard reserve solution, mL;

c ——the exact concentration of sodium thiosulfate solution, mol/L.

Have parallel titration twice. Error should be less than 0.05mL, otherwise recalibration is needed.

(3) Formaldehyde standard solution: before use, dilute the formaldehyde standard reserve solution with water and make 1.00mL solution contain $10\mu g$ formaldehyde. Immediately take 10.00mL solution, transfer it to 100mL volumetric flask, then add 5mL absorbant stock solution into the volumetric flask, and fix the volume with water to 100mL. 1.00mL of the solution contains $1.00\mu g$ formaldehyde. Place it for 30min before preparing standard chromophores. The standard solution can be stable for 24h.

4 Experimental procedure

4.1 Sampling point selection

According to the actual situation of the selected indoor environment, the site layout should be planned before sampling. Sampling point selection should follow the following principles.

(1) Number of sampling points: The number of sampling points is determined according to the size of the indoor area monitored and the on-site situation. In principle, rooms less than $50m^2$ should have 1-3 points; 3-5 points are set for 50-$100m^2$; At least 5 points are set above $100m^2$. Points are evenly distributed on diagonal or in plum pattern.

(2) The sampling point should avoid the air vent and be more than 0.5m away from the wall.

(3) Height of sampling point: In principle, the height of sampling point should be consistent with the height of human breathing, and the relative height should be 0.5-1.5m.

4.2 Sample collection

A large bubble absorption tube containing 5mL absorption liquid is used to collect 10L gas at a flow rate of 0.5L/min. Record the temperature and atmospheric pressure at the sampling point. Samples should be analyzed at room temperature within 24h after sampling.

4.3 Draw standard curves

Take 10mL colorimetric tubes with plug and prepare formaldehyde standard series with formaldehyde standard solution according to Table 4-1-5.

Chapter 4 Environmental chemistry experiments

Table 4-1-5 Formaldehyde standard series

Tube No.	0	1	2	3	4	5	6	7	8
Standard working fluid volume/mL	0.00	0.10	0.20	0.40	0.60	0.80	1.00	1.50	2.00
Absorbing liquid volume/mL	5.0	4.9	4.8	4.6	4.4	4.2	4.0	3.5	3.0
Formaldehyde weight/μg	0.0	0.1	0.2	0.4	0.6	0.8	1.0	1.5	2.0

In each tube, add 0.4mL of 10g/L ammonium ferric sulfate solution and shake well. Place for 20min. Water was used as a reference to determine the absorbance of each solution in a 1cm colorimeter at 630nm. The curve is drawn with formaldehyde weight as the x-coordinate and absorbance as the y-coordinate, and the regression slope is calculated. The slope reciprocal is taken as the calculation factor B_s (μg/absorbance) of the sample determination.

4.4 Sample determination

After sampling, transfer all the sample solution into the colorimetric tube, moisten the absorption tube with a small amount of absorption liquid, and merge to make the total volume 5mL. Determine the absorbance (A) according to the procedure of drawing the standard curve. At the same time, 5mL unsampled absorbent solution is used as reagent blank to determine the absorbance (A_0) of reagent blank.

5 Data Processing

(1) Convert the sampling volume into the sampling volume in the standard state according to the equation (4-1-3):

$$V_0 = V_t \times \frac{T_0}{273+t} \times \frac{P}{P_0} \quad (4\text{-}1\text{-}3)$$

Where V_0——sampling volume in the standard state, L;

V_t——sampling volume, i.e, sampling flow (L/min) × sampling time (min);

t——temperature at the sampling point, ℃;

T_0——thermodynamic temperature in the standard state, K;

P——atmospheric pressure at the sampling point, kPa;

P_0——atmospheric pressure in the standard condition, 101.3kPa.

(2) The formaldehyde concentration in the air is calculated as the equation (4-1-4):

$$\rho = \frac{(A-A_0)B_s}{V_0} \quad (4\text{-}1\text{-}4)$$

Where, ρ——formaldehyde concentration in air, mg/m^3;

A——absorbance of the sample solution;

A_0——absorbance of blank solution;

B_s——calculation factor calculated by experiment, μg/absorbance;

V_0——converted to the sampling volume in the standard state, L.

6 Questions

(1) Try to analyze and discuss the main sources of formaldehyde in indoor air.

(2) Understand other monitoring and analysis methods of formaldehyde. Compared with phenol reagent spectrophotometry, What are the advantages and disadvantages of other methods?

Experiment 4 Liquid phase oxidation simulation of SO_2 in ambient air

1 Experimental purpose

(1) Understand the liquid phase oxidation process of SO_2.

(2) Master the method of indirectly examining the liquid phase oxidation process of SO_2 by pH method.

2 Experimental principle

The liquid phase oxidation process of SO_2 is the main way of atmospheric precipitation acidification. First of all, SO_2 is dissolved in water and generate primary and secondary ionization with $SO_2 \cdot H_2O$, HSO_3^-, SO_3^{2-} and H^+. The distribution of dissolved sulfur is shown in the Fig. 4-1-1.

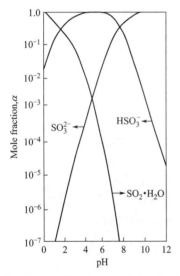

Fig. 4-1-1 Distribution of dissolved sulfur (IV)

The existence form of dissolved total sulfur is not only related to the concentration of SO_2, but also related to the liquid phase pH. General conditions, the pH of typical atmospheric droplets is 2-6. At this time, the main existence form of dissolved S (IV) is HSO_3^-, then the dissolved S (IV) is oxidized to S (VI). Common liquid oxidizers include O_2, O_3, H_2O_2 and free radicals. O_2 dissolved in the water is the most common and major oxidizer. During the oxidation of SO_2 by O_2, Fe (III) and Mn (II) can both have the effects of catalysts.

$$Mn^{2+} + SO_2 \Longrightarrow MnSO_2^{2+}$$
$$2MnSO_2^{2+} + O_2 \Longrightarrow 2MnSO_3^{2+}$$
$$MnSO_3^{2+} + H_2O \Longrightarrow Mn^{2+} + 2H^+ + SO_4^{2-}$$

Total reaction:
$$2SO_2 + 2H_2O + O_2 \Longrightarrow 2SO_4^{2-} + 4H^+$$

The Fe (III) and Mn (II) in the water mainly come from dust and other impurities in the atmosphere.

The S (IV) in atmospheric droplets mainly exists as the form of HSO_3^-. Therefore, in this experiment, Na_2SO_3 solution is substituted for droplets that absorb SO_2 to simulate and study liquid phase oxidation process of S (IV) under different conditions. Since H^+ concentration of the solution increases and pH decreases in the process of SO_3^{2-} being oxidized to

SO_4^{2-}, this experiment estimates the liquid phase oxidation rate of SO_2 by measuring the pH change of the solution. At the same time, different catalysts are added and their catalytic effects are compared. In this experiment, $MnSO_4$ is used to simulate Mn (II), $NH_4Fe(SO_4)_2$ is used to simulate Fe (III), and dust and coal ash are used to simulate the actual atmospheric dust and other impurities in the droplets.

3 Experimental instruments and reagentss

3.1 Instruments

(1) Precise pH meter for 2.
(2) Magnetic stirrers for 6.
(3) Small air pump.
(4) 2L beaker for 1.
(5) 250mL beakers for 6.
(6) 1L volumetric flasks for 3.

3.2 Reagents

(1) Sodium sulfite solution: 0.01mol/L. Dissolve 1.26g anhydrous Na_2SO_3 in water and fix volume to 1L.

(2) Manganese sulfate solution: 0.0005mol/L. Dissolved 0.141g anhydrous $MnSO_4$ in a beaker, adjust pH to 5 with dilute sulfuric acid, transfer it to a 1L volumetric flask, and dilute it to volume.

(3) Ammonium ferric sulfate solution: 0.0005mol/L. Add 0.241g $NH_4Fe(SO_4)_2 \cdot 12H_2O$ to a beaker, add a small amount of 1:4 dilute sulfuric acid and an appropriate amount of water, transfer it to a 1L volumetric flask, and dilute it to constant volume. Take an appropriate amount of solution with NaOH solution carefully adjusting pH to 5 (pay attention to avoiding precipitation).

(4) Dustfall-water suspension: Collect and take 0.2g atmospheric dustfall (can be taken from outdoor windowsills, etc.), put it into a 50mL beaker, add 30mL secondary water, stir, and adjust the pH value with dilute sulfuric acid to 5.

(5) Coal ash-water suspension: Weigh 0.1g coal ash, put it into a 50mL beaker, add 30mL secondary water, stir, and adjust the pH to 5 with dilute sulfuric acid.

(6) Diluted water: Take 1.5L secondary water in a 2L beaker, mix with air for 30min, stir with a magnetic stirrer, and finally adjust pH to 5 with dilute sulfuric acid.

(7) Dilute sulfuric acid solution: 0.01mol/L.

(8) Dilute sodium hydroxide solution: 0.01mol/L.

(9) Standard buffer solution: 0.05mol/L potassium hydrogen phthalate (pH=4.01) and 0.025mol/L KH_2PO_4-0.025mol/L Na_2HPO_4 (pH=6.86).

4 Experimental procedure

4.1 Preparation for simulation experiment

(1) Six 250mL beakers, numbered 1-6, are taken to simulate the six conditions of

adding no catalyst, adding manganese catalyst, adding iron catalyst, adding iron manganese catalyst, adding dustfall catalyst, and adding coal ash catalyst, respectively.

(2) Add 190mL diluted water into beaker No. 1-4, and 10mL 0.01mol/L Na_2SO_3 solution. Add 160mL diluted water into beaker No. 5 and beaker No. 6 respectively, and 10mL 0.01mol/L Na_2SO_3 solution.

(3) Quickly add the following reagents into the beaker No. 2-6 successively: No. 1, 2mL deionized water; No. 2, 2mL 0.0005mol/L $MnSO_4$ solution; No. 3, 2mL 0.0005mol/L $NH_4Fe(SO_4)_2$ solution; No. 4, 0.0005mol/L $MnSO_4$ solution and 0.0005mol/L $NH_4Fe(SO_4)_2$ solution are 1mL each; No. 5, 30mL dustfall-water suspension; No. 6, 30mL ash-water suspension.

(4) After adding various reagents, place six beakers on the magnetic stirrer for continuous stirring, and quickly adjust the pH of each solution to 5.0 with dilute H_2SO_4 and dilute NaOH solution, and start timing.

4.2 Liquid phase oxidation process

The change of solution pH in each beaker is measured and recorded at regular intervals (5, 10, 15, 20, 25, 30, 40, 50, 60, 70min).

5 Data processing

The curve of solution pH with time in each system is plotted with pH as the ordinate and time as the abscissa. Evaluate and compare the speed of oxidation reaction in different systems. Analyze and compare the catalytic action of each catalyst.

6 Questions

(1) Why can the liquid phase oxidation rate of SO_2 be estimated by the change of pH?
(2) What factors will affect the oxidation rate of SO_2?
(3) What is the key to the success of this experiment?

Experiment 5 Determination of SO_2 in air

1 Experimental purpose

(1) Understand the principle and method of sulfur dioxide colorimetry.
(2) Be familiar with and master collection technologies of air pollutant samples.

2 Experimental principle

Pararosaniline hydrochloride method for the determination of sulfur dioxide in the atmosphere is widely used at home and abroad. The method is sensitive and selective, and can be used for short time sampling (such as 20-30min) or long time sampling (such as 24h), but the absorbent is highly toxic.

After sulfur dioxide is absorbed by potassium mercuric tetrachloride solution, stable dichlorosulfite complex is formed, and then it is combined with formaldehyde and pararosaniline hydrochloride to form a purplish red complex, which is determined by colorimetry. The main interfering substances are nitrogen oxides, ozone, manganese, iron, chromium and so on. The addition of ammonium sulfamate can eliminate the interference of nitrogen oxide. Ozone can decompose by itself after being placed for a period of time after sampling. The addition of phosphoric acid and disodium salt of ethylenediamine tetraacetic acid can eliminate or reduce the interference of some heavy metals. Trace ammonia, sulfide and aldehydes in the atmosphere do not interfere. When the sampling volume is 30L, the minimum detected concentration of sulfur dioxide is $0.025 mg/m^3$. When the concentration is lower than $0.025 mg/m^3$, the sample volume should be increased, but the absorption efficiency of the sample must be checked and corrected. When sulfur dioxide concentration is higher than $0.025 mg/m^3$, the absorption efficiency is greater than 98%.

3 Experimental instruments and reagents

3.1 Instruments

(1) Absorption tube: Porous glass plate absorption tube, small impact absorption tube or large bubble absorption tube which is used for sampling from 30min to 1h. 125mL porous glass plate absorption bottle or 125mL air washer bottle for 24h sampling.

(2) Air sampler: Flow range is 0-1L/min.

(3) Spectrophotometer.

(4) Thermostatic water bath.

3.2 Reagents

(1) 0.04mol/L potassium mercuric tetrachloride (TCM) absorption solution: 10.9g mercuric chloride ($HgCl_2$), 6.0g potassium chloride (KCl) and 0.066g disodium salt of ethylenediamine tetraacetic acid (EDTA-2Na) are weighed. Dissolve them in water and dilute to 1L. This solution has a pH value of about 4, and the pH is adjusted to about 5.2 on the pH meter with 0.01mol/L sodium hydroxide solution. This reagent can be stabilized for 6 months in sealed container.

(2) 0.6% ammonium sulfamate solution: 0.6g ammonium sulfamate ($H_2NSO_3NH_4$) is dissolved in water. Dilute it to 100mL, and prepare temporarily.

(3) 0.2% formaldehyde solution: 1.4mL 36%-38% formaldehyde is dissolved in water, diluted to 250mL, stored in the refrigerator. It can be stable for one and a half months.

(4) 0.05mol/L iodine reserve solution: Weigh 12.7g iodine into a beaker, add 40g potassium iodide (KI), add 25mL water, stir until all reagents are dissolved, dilute with water to 1L and store it in a brown reagent bottle.

(5) 0.005mol/L iodine solution: 50mL 0.05mol/L iodine reserve solution is taken, diluted with water to 500mL and stored in a brown reagent bottle.

(6) Starch indicator: Weigh 0.2g soluble starch (0.4g zinc dichloride can be added to prevent corrosion), mix a small amount of water into a paste, pour into 100mL boiling water and continue to boil until the solution is clear. Cool and store it in reagent bottles.

(7) 0.015mol/L standard solution of potassium iodate: Weigh 3.2100g of potassium iodate (KIO_3, guaranteed reagent, baked at 110℃ for 2h), and dissolve it in water. Transfer solution to a 1000mL volumetric flask, and dilute it with water to the mark line.

(8) 0.1mol/L sodium thiosulfate storage solution: Weigh 25g sodium thiosulfate ($Na_2S_2O_3 \cdot 5H_2O$), dissolve it in 1L newly boiled but cooled water, add 0.2g anhydrous sodium carbonate, store the solution in a brown reagent bottle, and calibrate its concentration after one week. If the solution is cloudy, filter it.

Calibration method: after absorbing 10.00mL 0.1mol/L standard solution of potassium iodate, put it into a 250mL iodine measuring bottle, add 70mL newly boiled but cooled water, add 1g potassium iodide, shake the solution until reagents are completely dissolved, then add 3.5mL glacial acetic acid (or 10mL 1mol/L hydrochloric acid solution), immediately plug the bottle, and mix well. After being placed in the dark for 5min, 0.1mol/L sodium thiosulfate solution is used to titrate until pale yellow. Add 5mL newly prepared 0.2% starch indicator, and the solution appears blue at the same time. Then the solution is titrated until the blue just disappears. Calculate the concentration of sodium thiosulfate solution.

(9) 0.01mol/L sodium thiosulfate solution: 50.00mL calibrated 0.1mol/L sodium thiosulfate solution is placed in a 500mL volumetric flask and diluted to the mark line with newly boiled but cooled water.

(10) Sulfur dioxide standard solution: First prepare sodium sulfite solution. Weigh 0.200g sodium sulfite (Na_2SO_3) and 0.010g disodium salt of ethylenediamine tetraacetic acid, dissolve them in 200mL fresh boiled but cooled water, gently shake well (avoid oscillation to prevent oxygenation), and place for 2-3h for calibration. The solution is equivalent to 320-400μg sulfur dioxide per milliliter.

Calibration method is as follows:

Four 250mL iodine measuring bottles (A_1, A_2, B_1, B_2) are taken and 50.00mL 0.005mol/L iodine solution is added respectively. Add 25mL water into A_1 and A_2 respectively, add 25.00mL above-mentioned sodium sulfite solution into B_1 and B_2 respectively, and plug the bottle.

Immediately absorb 2.00mL of the above-mentioned sodium sulfite solution and add it to a 100mL volumetric flask with 40-50mL potassium mercuric tetrachloride solution to form stable dichlorosulfite complex.

Dilute the solution in 100mL volumetric flasks with potassium mercuric tetrachloride solution to the standard line and shake well.

After 4 bottles of A_1, A_2, B_1 and B_2 are placed in the dark for 5min, they are titrated to light yellow with 0.01mol/L sodium thiosulfate standard solution. 5mL newly prepared 0.2% starch indicator is added. The titrating is continued until the blue color just disappears. The difference in volume of the standard solution of sodium thiosulfate used for parallel titration shall not be more than 0.10mL.

The concentration of sulfur dioxide in the 100mL volumetric flask is as formula (4-1-5):

$$c(SO_2) = (A-B)N \times 32000 \times 2.00/(25.00 \times 100) \tag{4-1-5}$$

Where, c——concentration of sulfur dioxide solution, $\mu g/mL$;

A——mean volume value of the standard solution of sodium thiosulfate used for blank titration, mL;

B——mean volume value of the standard solution of sodium thiosulfate used for titration, mL;

N——concentration of sodium thiosulfate standard solution, mol/L.

Based on the calculated concentration of sulfur dioxide, the absorbent solution is then diluted to 2.0$\mu g/mL$ standard solution of sulfur dioxide. The solution is refrigerated at 0-4℃ and the concentration remains the same for a week.

(11) Purification of pararosaniline hydrochloride (pararosaniline): 500mL n-butanol and 500mL 1mol/L hydrochloric acid solution are taken, and the mixture is shaken well in a 1L separating funnel to achieve an equilibrium. Weigh 0.100g pararosaniline in a small beaker, add about 30mL balanced 1mol/L hydrochloric acid solution, stir it, and place it until completely dissolved. Wash it with 1mol/L hydrochloric acid for several times into a 250mL separating funnel, and the total volume of solution shall not exceed 50mL. Add 100mL balanced n-butanol, shake for a few minutes, stand until two-phase stratification, transfer the lower hydrochloric acid solution containing the pararosaniline into another separating funnel, add 100mL balanced n-butanol, and then extract. According to this operation, 50mL n-butanol is used for repeated extraction for 6 times. Retain the water phase, avoid loss as much as possible, and discard the organic phase. Finally, the water phase is filtered into a 50mL volumetric flask and diluted to the mark with 1mol/L hydrochloric acid solution. This pararosaniline reserve solution (concentration is 0.2%) is light brown and yellow, which should meet the following conditions:

The maximum absorption peak of pararosaniline reserve solution in acetic acid-sodium acetate buffer solution is found at 540nm. The solution for determination of absorption curve is prepared according to the following method: 1.00mL purified pararosaniline reserve solution is drawn into a 100mL volumetric flask. Dilute it with water to the standard line, and shake well. 5.00mL diluent is put into a 50mL volumetric flask, and 5.00mL 1mol/L acetic acid-sodium acetate buffer solution is added, diluted with water to the mark line, and the absorption curve is determined after 1h.

Reagent blank value is sensitive to temperature. When using pararosaniline using solution prepared by the storage solution to draw the standard curve at 22℃ according to this operation method, the absorbance of reagent blank solution measured at 548nm with 1cm cuvette should not exceed 0.170.

The slope of the standard curve drawn under the above conditions should be (0.030± 0.002) absorbance/μg SO_2.

(12) 0.016% pararosaniline using solution: Draw 20.00mL 0.2% pararosaniline reserve solution into a 250mL volumetric flask, add 25mL 3mol/L phosphoric acid solution, and dilute it to the mark line with water. Keep for at least 24h before use. The solution

stabilizes for more than 9 months.

(13) 1mol/L hydrochloric acid solution: 86mL concentrated hydrochloric acid (density is $1.19g/cm^3$) is taken and diluted to 1L with water.

(14) 3mol/L phosphoric acid solution: Take 41mL concentrated phosphoric acid (H_3PO_4, 85%) and dilute it with water to 200mL.

(15) 1mol/L acetic acid-sodium acetate buffer solution: Weigh 13.6g sodium acetate ($NaCH_3COO \cdot 3H_2O$), dissolve it in water, transfer it to a 100mL volumetric flask, add 5.7mL glacial acetic acid, and dilute it with water to marked line. The solution has the pH of 4.7.

The water for the reagents is distilled water without oxidants.

4 Experimental procedure

4.1 Sample collection

When the sampling time is 30min or 60min, 10mL absorbent solution is used and the flow rate is 0.5L/min. When the average concentration of 24h is measured, 75-100mL absorbent solution is used to continuously sample at the flow rate of 0.2-0.3L/min for 24h.

4.2 Standard curve drawing

Seven 25mL volumetric flasks are taken and standard curves are prepared according to the following table 4-1-6.

Table 4-1-6   standard curves

Bottle No.	0	1	2	3	4	5	6
Sulfur dioxide standard solution ($2\mu g/mL$) volume/mL	0.00	0.50	2.00	4.00	6.00	8.00	10.00
Potassium mercuric tetrachloride solution volume/mL	10.00	9.50	8.00	6.00	4.00	2.00	0.00
Sulfur dioxide weight/μg	0.0	1.0	4.0	8.0	12.0	16.6	20.0

Add 1.00mL 0.6% ammonium sulfamate solution in each of the above bottles, and shake well. Add 2.00mL 0.2% formaldehyde solution and 5.00mL 0.016% pararosaniline using solution, dilute it with newly boiled and cooled water to marked line, and shake well. When the room temperature is 15-20℃, the color is developed for 30min. When the room temperature is 20-25℃, the color is developed for 20min. The color is developed for 15min at the room temperature of 25-30℃. The absorbance is measured at 548nm with a 1cm cuvette and water is the reference. (Since reagent blank is sensitive to temperature and is easily affected by the temperature of the spectrophotometer, water is used as the reference).

In order to improve accuracy, constant temperature water bath can be used. The difference between the temperature of drawing the standard curve and the temperature of measuring the sample should not exceed ±2℃.

The regression equation of standard curve is calculated by least square method.

$$Y = bX + a \tag{4-1-6}$$

Where, Y——$A - A_0$ is the difference between standard solution absorbance (A) and reagent blank solution absorbance (A_0);

X——sulfur dioxide weight, μg;

B——slope of the regression equation;

A——intercept of the regression equation.

4.3 Sample determination

If there are impurities in the sample, they should be removed by centrifugation.

For samples whose sampling time is 30min or 1h, the absorbent can be transferred into 25mL volumetric flasks and the absorbent tubes can be rinsed with about 5mL water. When measuring the average concentration of 24h, adjust the sample volume to 75mL or 100mL marked line with absorbent, and draw 10.00mL sample solution into a 25mL volumetric flask.

The sample is placed for 20min to decompose the ozone.

For each batch of samples, reagent blank solution and control samples should be measured to check the reliability and operational accuracy of the reagents. Preparation method: 10.00mL potassium mercuric tetrachloride solution is collected in a 25mL volumetric flask and prepared for reagent blank solution. 2.00mL standard solution of sulfur dioxide (2.0μg/mL) is collected in a 25mL volumetric flask and 8.00mL potassium mercuric tetrachloride solution is added to prepare for the control sample.

In the reagent blank solution, control sample and all samples, 1.00mL 0.6% ammonium sulfamate solution is added. Shake well and place for 10min to remove the interference of nitrogen oxides. The following steps are the same as the standard curve drawing.

If the difference between the temperature of measuring the sample and that of drawing the standard curve is no more than 2℃, the difference between their reagent blank absorbance should not exceed 0.03. If this value exceeds, the standard curve should be redrawn.

If the absorbance of the sample is between 1.0 and 2.0, it can be diluted with reagent blank solution, and the absorbance can be measured after several minutes until the measured absorbance value is between 0.03 and 1.0. The dilution factor should not be more than 6 times.

5 Data processing

$$c(SO_2) = \frac{[(A-A_0)-a]D}{bV_r} \qquad (4\text{-}1\text{-}7)$$

Where, c——concentration of sulfur dioxide solution, mg/m^3;

A——absorbance of sample solution;

A_0——absorbance of reagent blank solution;

b——slope of the regression equation;

a——intercept of regression equation;

V_r——sampling volume converted to reference state(25℃, $1.01×10^5$Pa), L;

D——dilution factor(for 30min or 1h sample, $D=1$, for 24h mean concentration, $D=$ 7.5 or 10).

6 Notes

(1) Temperature has an effect on color development. The higher the temperature, the larger the blank value. When temperature is high, the color will develop faster and fade faster. So it's better to use constant temperature water bath to control the temperature of color development and decide color development temperature and time according to room temperature.

(2) Purification of pararosaniline can reduce the absorbance of reagent blank and improve the sensitivity of the method. Although increasing acidity can also reduce the absorbance of reagent blank, the sensitivity of the method also decreases.

(3) Because hexavalent chromium can fade the purplish red complex and cause negative interference, it is necessary to avoid washing glassware with sulfuric acid-chromic acid lotion. If it has been washed with sulfuric acid-chromic acid lotion, you need soak and wash it with (1+1) hydrochloric acid solution, and then wash with water fully to wash hexavalent chromium out.

(4) Used volumetric flasks and cuvettes should be timely washed with acid, otherwise it is difficult to wash out red. The volumetric flask is washed with (1+4) hydrochloric acid solution, and the cuvette is washed with mixture of (1+4) hydrochloric acid and 1/3 volume ethanol.

(5) Potassium mercuric tetrachloride solution is highly toxic. Please be careful when using it. If it is splashed on skin, rinse it immediately with water.

Treatment method of waste liquid containing potassium mercuric tetrachloride: add about 10g sodium carbonate to neutral in every liter of waste liquid, add 10g zinc granules, stir for 24h under black cloth, pour supernatant into glass tank, and drop saturated sodium sulfide solution until no precipitation is generated. Discard the solution and transfer the precipitate to an appropriate container. This method can remove 99% of the mercury in the waste liquid.

(6) When preparing sodium sulfite solution, a small amount of disodium EDTA should be added. When SO_3^{2-} is oxidized to SO_4^{2-} by dissolved oxygen in water, the concentration of SO_3^{2-} is relatively stable with the addition of disodium EDTA. The coordination reaction is catalyzed by reagent and trace Fe^{3+} in water.

Experiment 6 Determination of nitrogen oxides in air with N-(1-naphthyl)ethylenediamine dihydrochloride spectrophotometry

Common nitrogen oxides include nitric oxide (NO, colorless), nitrogen dioxide (NO_2, reddish brown), nitrous oxide (N_2O), dinitrogen pentoxide (N_2O_5), etc. Except dinitrogen pentoxide, which is normally solid, other nitrogen oxides are normally gaseous. As air pollutants, nitrogen oxides (NO_x) often refer to NO and NO_2.

The natural emission of NO_x mainly comes from the decomposition of organic matter in soil and ocean, which belongs to the natural nitrogen cycle process. The majority of NO_x emissions from human activities come from the combustion process of fossil fuels, such as the fuel combustion in automobiles, airplanes, internal combustion engines, and industrial kilns. They also come from the process of producing and using nitric acid, such as nitrogen fertilizer plants, organic intermediate plants, non-ferrous and ferrous metals smelters, etc. According to estimates in the early 1980s, approximately 53 million tons of NO_x were emitted into the atmosphere annually by human activities worldwide. NO_x is extremely harmful to the environment. It is not only one of the main substances that form acid rain, but also an important substance that forms photochemical smog in the atmosphere and an important factor that consumes O_3. Nitrogen oxides can stimulate the lungs and make people more difficult to resist respiratory system disease such as colds. People with respiratory system problems, such as asthma patients, are more vulnerable to nitrogen dioxide.

According to Chinese "Ambient Air Quality Standard", the environmental standard for nitrogen oxides is divided into two levels. The first level standard is consist with the second level standand. The daily average maximum concentration limit is $100\mu g/m^3$. It is applicable to nature reserves, scenic spots and other areas requiring special protection, as well as residential areas, commercial traffic and residential mixed areas, cultural areas, industrial areas, and rural areas.

Therefore, mastering the determination of nitrogen oxide content in the air is of great significance.

1 Experimental purpose

(1) Master the basic principle and operational steps of Saltzman method for determining nitrogen oxide content in air.

(2) Understand the impact of nitrogen oxide concentration levels in ambient air on air quality.

2 Experimental principle

Nitrogen oxides refer to the oxides of nitrogen (calculated as NO_2) present in the form of nitric oxide and nitrogen dioxide in the air. Saltzman experimental coefficient refers to the ratio of the amount of azo dye generated by the absorption solution during gas production, equivalent to the amount of nitrite ions to the total amount of nitrogen dioxide passing through the sampling system with the nitrogen dioxide calibration gas mixture prepared by the permeation method. The oxidation coefficient refers to the ratio of the amount of nitric oxide in the air oxidized to nitrogen dioxide and absorbed by the absorption solution to generate azo dye after passing through the acid potassium permanganate solution oxidation tube to the total amount of nitric oxide passing through the sampling system.

Nitrogen dioxide in the air is absorbed by the absorption solution in the first absorption bottle in series and reacts to form pink azo dye. The nitric oxide in the air does not react with the absorption solution. When passing through the oxidation tube, it is oxidized by acid potassium permanganate solution to nitrogen dioxide, which is absorbed by the absorption solution in the second absorption bottle in series and reacts to generate pink azo dye. The absorbance of the generated azo dye at 540nm is proportional to the content of nitrogen dioxide. Measure the absorbance of the samples in the first and second absorption bottles respectively, calculate the mass concentrations of nitrogen dioxide and nitric oxide in the two absorption bottles, and the sum of the two is the mass concentration of nitrogen oxides (calculated as NO_2).

3 Experimental instruments and reagents

3.1 Instruments

(1) Spectrophotometer.

(2) Air samplers: Flow range of the portable air sampler is 0.1-1.0L/min. When the sampling flow rate is 0.4L/min, the relative error is less than ±5%. For thermostatic and semi-automatic continuous air sampler, when the sampling flow rate is 0.2L/min, the Xrelative error is less than ±5%, and the absorption liquid temperature can be maintained at (20±4)℃. The sampling connection pipeline is a borosilicate glass tube, stainless steel tube, polytetrafluoroethylene tube, or silicone tube. Its inner diameter is approximately 6mm. It shall be as short as possible and in no case more than 2m. It is equipped with a downward facing air inlet;

(3) Absorption bottles: Porous glass plate absorption bottles capable of holding 10mL, 25mL, or 50mL absorption solution, with a liquid column not lower than 80mm. The glass plate resistance, uniformity of bubble dispersion, and sampling efficiency of the absorption bottle shall be checked according to Appendix A of "Ambient air—Determination of nitrogen oxides—Saltzman method" (GB/T 15436—1995). Fig. 4-1-2 shows two more suitable porous glass plate absorption bottles. Use a brown absorption bottle or cover a black light shield on the absorption bottle during the sampling process. New porous glass absorption bottles or used porous glass absorption bottles should be soaked in (1+1) HCl for more than 24hours and washed with clean water.

(4) Oxidation bottle: Gas bottles which can hold 5mL, 10mL or 50mL acid potassium permanganate solution, and the liquid column height shall not be less than 80mm. After use, soak and wash bottles with hydroxylammonium chloride solution. Fig. 4-1-3 shows two more suitable oxidation bottles.

3.2 Reagents

Unless otherwise specified, analytical pure reagents conforming to national standards or professional standards, distilled water without nitrite ions, deionized water or water of equivalent purity shall be used for analysis. If necessary, 0.5g potassium permanganate

($KMnO_4$) and 0.5g barium hydroxide [$Ba(OH)_2$] can be added to each liter of water in an all glass distiller for redistillation.

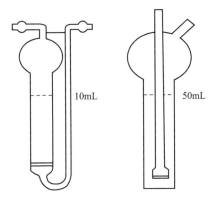

Fig. 4-1-2 Schematic diagram of porous glass plate absorption bottles

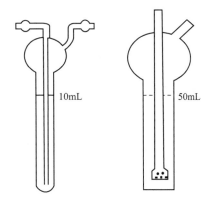

Fig. 4-1-3 Schematic diagram of oxidation bottles

(1) Glacial acetic acid.

(2) Hydroxylamine hydrochloride, $\rho = 0.2$-0.5g/L.

(3) Sulfuric acid solution, $c(1/2 H_2SO_4) = 1$mol/L. Take 15mL concentrated sulfuric acid ($\rho = 1.84$g/mL), slowly add it to 500mL water, stir well, and cool for later use.

(4) Acid potassium permanganate solution, $\rho(KMnO_4) = 25$g/L. Weigh 25g potassium permanganate into a 1000mL beaker, add 500mL water, heat slightly to dissolve it completely, then add 500mL 1mol/L sulfuric acid solution, stir it evenly, and store it in a brown reagent bottle.

(5) N-(1-naphthyl)ethylenediamine hydrochloride [$C_{10}H_7NH(CH_2)_2NH_2 \cdot 2HCl$] stock solution, $\rho = 1.00$g/L. Weigh 0.50g N-(1-naphthyl)ethylenediamine hydrochloride into a 500mL volumetric flask, dissolve it with water and dilute the solution to the scale. This solution is stored in a closed brown bottle and refrigerated in a refrigerator, which can be stably stored for three months.

(6) Chromogenic solution: Weigh 5.0g sulfanilic acid ($NH_2C_6H_4SO_3H$) and dissolve it in about 200mL 40-50℃ water, cool the solution to room temperature, transfer all of it into a 1000mL volumetric flask, add 50mL N-(1-naphthyl)ethylenediamine hydrochloride stock solution and 50mL glacial acetic acid, and dilute with water to the scale. This solution is stored in a closed brown bottle and can be stable for three months when stored in a dark place below 25℃. If the solution appears light red, it should be discarded and reprepared.

(7) Absorption liquid. When using, mix the chromogenic solution and water in a ratio of 4:1 (volume fraction) to form the absorption solution. The absorbance of the absorption solution should be less than or equal to 0.005.

(8) Nitrite (NO_2^-) standard stock solution, $\rho = 250\mu$g/mL. Accurately weigh 0.3750g sodium nitrite ($NaNO_2$, guaranteed reagent, dried at 105℃±5℃ before use to constant weight), dissolve it in water, transfer it into a 1000mL volumetric flask, and dilute it with

water to the mark. This solution is stored in a closed brown bottle in a dark place and can be stably stored for three months;

(9) Nitrite (NO_2^-) standard working solution, $\rho = 2.5\mu g/mL$. Accurately transfer 1.00mL nitrite standard stock solution into a 100mL volumetric flask, and dilute it with water to the mark. Prepare when needed.

4 Experimental procedure

4.1 Sampling

Take two porous glass plate absorption bottles with 10.0mL absorption solution and one oxidation bottle with 5-10mL acid potassium permanganate solution (the liquid column height is not less than 80mm), use the shortest silicone tube to connect the oxidation bottle in series between the two absorption bottles, and collect gas 4-24L at a flow rate of 0.4L/min. Sample collection, transportation, and storage should be kept away from light. Samples should be analyzed as soon as possible after collection.

4.2 Drawing standard curves

Take six 10mL colorimetric tubes with plugs and prepare nitrite standard solution according to Table 4-1-7. Take the corresponding volume of sodium nitrite standard working solution according to Table 4-1-7, add water to 2.00mL, and add 8.00mL chromogenic solution.

Table 4-1-7 NO_2^- standard solution

Tube No.	0	1	2	3	4	5
Standard working fluid volume/mL	0.00	0.40	0.80	1.20	1.60	2.00
Water volume/mL	2.00	1.60	1.20	0.80	0.40	0.00
Chromogenic solution volume/mL	8.00	8.00	8.00	8.00	8.00	8.00
NO_2^- mass concentration/($\mu g/mL$)	0.00	0.10	0.20	0.30	0.40	0.50

Mix each tube well and place them in a dark place for 20min (more than 40min at room temperature below 20℃). Measure the absorbance using a 10mm cuvette at the wavelength of 540nm, with water as the reference. Deduct the absorbance of tube 0 and get the corresponding mass concentration ($\mu g/mL$) of NO_2^-. Use the least square method to calculate the regression equation of the standard curve.

Control the slope of the standard curve between 0.960 and 0.978 absorbance • mL/μg. Control intercept between 0.000 and 0.005 (when drawing a standard curve with the volume of 5mL, the slope of the standard curve should be controlled between 0.180 and 0.195 absorbance • mL/μg, and the intercept should be controlled within ± 0.003).

Blank test: Take the blank absorption solution without sampling in the laboratory, use a 10mm colorimetric dish, and measure the absorbance with water as the reference at the wavelength of 540nm.

4.3 Determination

After sampling, let it stand for 20min. When the room temperature is below 20℃, let it stand for more than 40min. Use water to replenish the absorption solution in the sampling bottle to the mark, and mix well. Measure the absorbance using a 10mm cuvette at the wavelength of 540nm, with water as the reference. Meanwhile, measure the absorbance of the blank sample. If the absorbance of the sample exceeds the upper limit of the standard curve, the sample should be diluted with the laboratory blank solution before measuring its absorbance. The dilution ratio shall not exceed 6.

5 Data processing

(1) Calculate the mass concentration of nitrogen dioxide ρ_{NO_2} (mg/m^3) in the air according to equation (4-1-8):

$$\rho_{NO_2} = \frac{(A_1 - A_0 - a)VD}{bfV_0} \quad (4\text{-}1\text{-}8)$$

(2) Mass concentration of nitric oxide ρ_{NO} (mg/m^3) in the air is calculated as nitric oxide (NO) according to equation (4-1-9):

$$\rho_{NO} = \frac{(A_2 - A_0 - a)VD}{bfV_0 K} \quad (4\text{-}1\text{-}9)$$

(3) Mass concentration of nitrogen oxides ρ_{NO_x} (mg/m^3) in the air is calculated as nitrogen dioxide (NO$_2$) according to equation (4-1-10):

$$\rho_{NO_x} = \rho_{NO_2} + \rho_{NO} \quad (4\text{-}1\text{-}10)$$

Where, A_1, A_2 —— absorbance of samples in the first and second absorption bottles connected in series;

A_0 —— absorbance of laboratory blank;

b —— slope of standard curve;

a —— intercept of standard curve;

V —— volume of absorption solution for sampling;

V_0 —— converted to sampling volume under standard state (101.325kPa, 273 K), L;

K —— NO→NO$_2$ oxidation coefficient, 0.68;

D —— dilution ratio of the sample;

f —— Saltzman experimental coefficient, 0.88 (when the mass concentration of nitrogen dioxide in the air is higher than 0.72mg/m^3, the value of f is 0.77).

6 Questions

(1) Analyze and discuss the main sources of nitrogen oxides in ambient air.

(2) What are the interference when measuring nitrogen oxides in the air? How to eliminate it?

Experiment 7　Determination of ozone in ambient air using sodium indigo disulfonate spectrophotometry

Ozone is an allotropy of oxygen, chemical formula is O_3, molecular weight is 47.998, gas is light blue, liquid is dark blue, and solid is purple black. The odor is similar to fishy odor, but when the concentration is too high, the odor is similar to chlorine gas. Ozone has strong oxidizing properties and is a stronger oxidant than oxygen. It can undergo oxidation reactions at lower temperatures. In chemical production, ozone oxidation can replace many catalytic or high-temperature oxidation methods, simplify production processes, and improve production efficiency. Liquid ozone can also be used as an oxidant for rocket fuel. Ozone exists in the atmosphere, with concentration ranging from 0.001 to $0.03 \mu g/mL$ near the earth's surface. This is due to the absorption of ultraviolet radiation with the wavelength of less than 185nm by oxygen in the atmosphere. The ozone layer can absorb harmful short wave (below 300nm) light from the sun, prevent it from reaching the ground, and protect organisms from UV damage.

The olfactory threshold of ozone is $0.02 mL/m^3$. If the concentration reaches $0.1 mL/m^3$, it will stimulate the mucosa, and if the concentration reaches $2 mL/m^3$, it will cause central nervous system disorders. The national environmental air quality standards (NAAQS) propose that the maximum acceptable concentration of ozone for humans within one hour is $260 \mu g/m^3$. Activity in an ozone environment of $320 \mu g/m^3$ for 1hour can cause coughing, difficulty breathing, and decreased lung function. Ozone can also participate in the reaction of unsaturated fatty acid, amino and other proteins in the organism, causing fatigue, cough, chest tightness, chest pain and other symptoms of people who have been directly exposed to high concentration of ozone for a long time.

In 1951, A.J. Hagen first pointed out that ozone (O_3) is a mixture of nitrogen oxides, hydrocarbons, and air formed through photochemical reactions. Later, F.W. Winter found that the chemical reaction products of O_3 and unsaturated hydrocarbons (such as hydrocarbons in automobile exhaust) had the same harmful effect as Los Angeles smog. Therefore, the increase of O_3 concentrations is a sign of photochemical smog pollution. The World Health Organization has taken the concentration level of ozone as one of the standards to judge the quality of the atmospheric environment, and issued an alarm of photochemical smog accordingly.

Therefore, mastering the determination of ozone in ambient air is of great significance.

1　Experimental purpose

(1) Master the basic principle and operational steps of determining ozone in ambient air using sodium indigo disulfonate spectrophotometry.

(2) Understand the impact of ozone concentration levels in ambient air on air quality.

2　Experimental principle

In the presence of phosphate buffer solution, ozone in the air reacts with the blue sodium indigo disulfonate in the absorption solution to fade and form sodium isatin disulfonate sodium. The absorbance is measured at 610nm, and the concentration of ozone in the air is quantified based on the degree of blue fading.

3　Experimental instruments and reagents

3.1　Instruments

(1) Spectrophotometer: Equipped with 20mm cuvettes, capable of measuring absorbance at the wavelength of 610nm.

(2) Air samplers: Flow range is 0.0-1.0L/min, flow rate is stable. When using, use a soap film flowmeter to calibrate the flow rate of the sampling system before and after sampling, and the relative error should be less than ±5%.

(3) Porous glass plate absorption tube: Filled with 10mL absorption solution inside and collect gas at the flow rate of 0.50L/min. The glass plate resistance should be 4-5kPa, and the bubbles should be evenly dispersed.

(4) Colorimetric tubes with plug: 10mL.

(5) Biochemical incubator or constant temperature water bath: Temperature control accuracy is ±1℃.

(6) Mercury thermometer: Accuracy is ±0.5℃.

3.2　Reagents

Unless otherwise specified, all reagents used in this experiment are analytical pure or chemical reagents that meet national standards, and the experimental water is freshly prepared deionized water or distilled water.

(1) Potassium bromate standard stock solution, $c(1/6KBrO_3)=0.1000$mol/L. Accurately weigh 1.3918g potassium bromide (guaranteed reagent, baked at 180℃ for 2h), place it in a beaker, add a small amount of water to dissolve it, transfer it into a 500mL volumetric flask, and dilute it with water to the mark.

(2) Potassium bromate-potassium bromide standard solution, $c(1/6KBrO_3)=0.0100$mol/L. Transfer 10.00mL potassium bromate standard stock solution into a 100mL volumetric flask, add 1.0g potassium bromide (KBr), and dilute it with water to the mark.

(3) Sodium thiosulfate standard stock solution, $c(Na_2S_2O_3)=0.1000$mol/L.

(4) Sodium thiosulfate standard working solution, $c(Na_2S_2O_3)=0.0050$mol/L. Before use, take sodium thiosulfate standard stock solution and dilute it accurately 20 times with newly boiled but cooled to room temperature water.

(5) Sulfuric acid solution, 1+6 (volume ratio).

(6) Starch indicator solution, $\rho=2.0$g/L. Weigh 0.20g soluble starch, mix with a small amount of water to form a paste, slowly pour in 100mL boiling water, and boil it until the solution is clear.

(7) Phosphate buffer solution, $c(KH_2PO_4\text{-}Na_2HPO_4)=0.050$ mol/L. Weigh 6.8g monopotassium phosphate (KH_2PO_4) and 7.1g anhydrous disodium hydrogen phosphate (Na_2HPO_4), dissolve them in water and dilute the solution to 1000mL;

(8) Sodium indigo disulfonate ($C_{16}H_{18}Na_2O_8S_2$) (abbreviated as IDS), Analytical, chemical, or biochemical reagent.

(9) IDS standard stock solution: Weigh 0.25g sodium indigo disulfonate, dissolve it in water, transfer it into a 500mL brown volumetric flask, dilute it with water to the mark, shake it well, store it in a dark place at room temperature for 24h, and calibrate it. This solution can be stable for 2 weeks when stored in a dark place below 20℃.

Calibration method: Accurately transfer 20.00mL IDS standard stock solution into a 250mL iodine volumetric flask, add 20.00mL potassium bromate-potassium bromide solution, add 50mL water, cover the bottle stopper, place it in a (16±1)℃ biochemical incubator (or water bath) until temperature of the solution is balanced with the temperature of water bath, add 5.0mL sulfuric acid solution, immediately plug, mix and start timing, place it in a dark place at (16±1)℃ for (35±1.0)min, add 1.0g potassium iodide and immediately plug it, shake it gently until dissolved, place it in the dark for 5min, titrate it with sodium thiosulfate solution until the brown color just fades to light yellow, add 5mL starch indicator solution, continue titrating until the blue color fades, and the end point is bright yellow. Record the consumed volume of sodium thiosulfate standard working solution.

The mass concentration of ozone ρ (μg/mL) is calculated from equation (4-1-11):

$$\rho = \frac{c_1V_1 - c_2V_2}{V} \times 12.00 \times 10^3 \qquad (4\text{-}1\text{-}11)$$

Where, ρ——Mass concentration of ozone, μg/mL;

c_1—— concentration of potassium bromate-potassium bromide standard solution, mol/L;

V_1—— volume of potassium bromate-potassium bromide standard solution added, mL;

c_2—— concentration of sodium thiosulfate standard solution used in titration, mol/L;

V_2——volume of sodium thiosulfate standard solution used in titration, mL;

V——volume of IDS standard stock solution, mL;

12.00——molar mass of ozone (1/4 O_3), g/mol.

(10) IDS standard working solution: Dilute the calibrated IDS standard stock solution step by step with phosphate buffer solution to IDS standard working solution. 1mL IDS standard working solution is equivalent to 1.00μg ozone. This solution can be stable for 1 week when stored in a dark place below 20℃.

(11) IDS absorption solution: Take an appropriate amount of IDS standard stock solution and dilute it with phosphate buffer solution according to the concentration of ozone in the air to form IDS absorption solution. 1mL IDS absorption solution is equivalent to 2.5μg (or 5.0μg) ozone. This solution can be stable for 1 month when stored in a dark place below 20℃.

4 Experimental procedure

4.1 Sample collection and preservation

Use a porous glass plate absorption tube with (10.00 ± 0.02) mL IDS absorption solution inside, cover it with a black light proof cover, and extract 5~30L gas at a flow rate of 0.5L/min. When the absorption solution fades by about 60% (compared to the blank sample on site), sampling should be stopped immediately. Samples should be strictly protected from light during transportation and storage. When it is confirmed that the mass concentration of ozone in the air is low and will not penetrate, a brown glass plate absorption tube can be used for sampling. The sample can be stable for at least 3 days when stored in a dark place at room temperature. Using the same batch of prepared IDS absorption solution, load it into a porous glass plate absorption tube and take it to the sampling site. Except for not collecting air samples, other environmental conditions remain the same as the sampling tubes used to collect air.

4.2 Standard curve drawing

Take 6 10mL colorimetric tubes with plugs and prepare standard solution according to Table 4-1-8.

Table 4-1-8 Standard solution

Test tube number	1	2	3	4	5	6
IDS standard solution volume/mL	10.00	8.00	6.00	4.00	2.00	0.00
Phosphate buffer solution volume/mL	0.00	2.00	4.00	6.00	8.00	10.00
Ozone mass concentration/(μg/mL)	0.00	0.20	0.40	0.60	0.80	1.00

Shake each tube well, use a 20mm colorimetric dish with water as a reference, and measure the absorbance at the wavelength of 610nm. Calculate the regression equation of the calibration curve using the least square method. The difference between the absorbance (A_0) of the zero concentration tube in the calibration series and the absorbance (A) of each standard solution tube is the ordinate and the ozone mass concentration is the abscissa.

$$y = aX + b \tag{4-1-12}$$

Where, y——$A_0 - A$, the difference between the absorbance of the blank sample and the absorbance of each standard solution tube;

X——ozone mass concentration, μg/mL;

b——the slope of the regression equation;

a——intercept of regression equation.

4.3 Sample determination

After sampling, connect a glass tip in series at the air inlet of the absorption tube, press the rubber pipette bulb at the air outlet of the absorption tube to transfer the sample solution in the absorption tube into a 25mL (or 50mL) volumetric flask, and wash the absorption tube with water several times to make the total volume is 25.0mL (or 50.0mL).

Measure absorbance at the wavelength of 610nm using a 20mm cuvette and with water as a reference.

5 Data processing

Calculate the mass concentration of ozone in the air according to equation (4-1-13):

$$\rho(O_3) = \frac{(A_0 - A - a)V}{bV_0} \qquad (4\text{-}1\text{-}13)$$

Where, $\rho(O_3)$——mass concentration of ozone in the air;
A_0——average absorbance of on-site blank samples;
A——sample absorbance;
b——slope of standard curve;
a——intercept of standard curve;
V——total volume of sample solution;
V_0——sampling volume converted to standard state (101.325kPa, 273 K).

The results are accurate to three decimal places.

6 Questions

(1) Analyze and discuss the main sources of ozone in the ambient air.

(2) Will nitrogen dioxide and sulfur dioxide in the air interfere with the determination of ozone?

Section 2 Water environmental chemistry experiments

Experiment 8 Determination of residual chlorine in drinking water (iodimetric titration)

When liquid chlorine is used as disinfectant in the disinfection process of drinking water, the residual chlorine in water is called residual chlorine after the interaction of liquid chlorine with reducing substances or microorganisms such as bacteria. It includes free residual chlorine and combined residual chlorine.

China's drinking water factories require free residual chlorine in finished wate is greater than 0.3mg/L and which is greater than 0.05mg/L in distribution system. In this experiment, residual chlorine in water is determined by iodimetry.

1 Experimental purpose

(1) Understand the significance of determination of residual chlorine in water.

(2) Master the principle and operation of determination of residual chlorine by iodimetry.

2 Experimental principle

Residual chlorine in the water reacts with KI in acidic solution to release isostoichiometric iodine, which is titrated by Na_2SO_3 standard solution with starch as indicator until the blue color disappears. The residual chlorine in the water is determined by the amount and concentration of standard solution consumed. The main reactions are as follows:

$$ClO^- + 2I^- + 2H^+ = I_2 + Cl^- + H_2O$$
$$I_2 + 2S_2O_3^{2-} = 2I^- + S_4O_6^{2-}$$

The measured value of this method is total residual chlorine, including HClO, ClO^-, NH_2Cl and $NHCl_2$.

3 Experimental instruments and reagents

3.1 Instruments

Iodine flask 300mL, burette 25mL, titration table, volumetric flask 500mL, beaker 200mL, glass rod, dropper, measuring cylinder 10mL.

3.2 Reagents

(1) Potassium iodide: Contain no free iodine and potassium iodate.

(2) (1+5) sulfuric acid solution.

(3) Standard solution of potassium dichromate, $c(1/6\ K_2Cr_2O_7) = 0.0250$ mol/L. Weigh 1.2259g potassium dichromate (guarantee reagent) (dried for 2h at 120℃ in advance and cooled in a dryer), dissolve it in water, transfer it to a 1000mL volumetric flask and dilute it with water to the mark line.

(4) Standard solution of sodium thiosulfate, $c(Na_2S_2O_3) = 0.05$ mol/L. Weigh 12.5g sodium thiosulfate ($Na_2S_2O_3 \cdot 5H_2O$), dissolve it in boiled and cooled water, and dilute it to 1000mL. Add 0.2g sodium carbonate and several grains of mercury iodide, and store it in a brown bottle. The solution can be stored for several months.

The calibration of the sodium thiosulfate standard solution: Transfer 20.00mL potassium dichromate standard solution into an iodine volumetric flask, add 50mL water and 1g potassium iodide, add 5mL (1+5) sulfuric acid solution, and stand for 5min. Use sodium thiosulfate standard solution to be calibrated to titrate the solution to pale yellow, add 1mL 1% starch solution, and continue titrating to blue disappeares (Note: the solution should be light green at this time, because it contains Cr^{3+}). Record the dosage of the Na_2SO_3 standard solution.

The concentration of standard solution of sodium thiosulfate is calculated as follows:

$$c = \frac{c_1 \times 20.00}{V} \qquad (4\text{-}2\text{-}1)$$

Where, c——concentration of sodium thiosulfate standard solution, mol/L;

c_1——concentration of potassium dichromate standard solution, mol/L;

20.00——volume of potassium dichromate solution sucked, mL;

V——volume of sodium thiosulfate standard solution to be calibrated, mL.

(5) Standard titration solution of sodium thiosulfate, $c(Na_2S_2O_3) = 0.0100$ mol/L. Transfer 100mL above calibrated standard solution of sodium thiosulfate of 0.05mol/L, move it into a 500mL volumetric flask, and dilute it to the scale with boiled but cooled water.

(6) 1% starch solution: Weigh 1.0g soluble starch, make it a paste with a small amount of distilled water, and add boiling distilled water to 100mL and mix well. After cooling, 0.1g salicylic acid or 0.4g zinc chloride are added as preservatives to prevent indicator spoilage.

(7) Acetate buffer solution (pH=4): Weigh 146g anhydrous sodium acetate, dissolve it in water, add 457mL acetic acid, and dilute it to 1000mL with water.

4 Experimental procedure

(1) Transfer 100mL water sample (200mL water sample is recommended if the concentration of residual chlorine in the water sample is less than 1mg/L) into a 300mL iodine flask and add 0.5g KI and 5mL acetate buffer solution (adjust pH≈4).

(2) Use 0.0100mol/L $Na_2S_2O_3$ standard titration solution to titrate the water sample until it turns pale yellow, add 1mL starch solution, continue titrating until the blue color disappears, and record the amount of $Na_2S_2O_3$ standard solution.

5 Data Processing

Residual chlorine in water is calculated as follows:

$$c(Cl_2) = \frac{c_1 V_1 35.453 \times 1000}{V_{water}} \qquad (4\text{-}2\text{-}2)$$

Where, c——mass concentration of total residual chlorine, mg/L;

c_1——concentration of sodium thiosulfate standard titration solution, mol/L;

V_1——volume of sodium thiosulfate standard titration solution, mL;

V_{water}——volume of water sample, mL;

35.46——molar mass of total residual chlorine (Cl_2), g/mol.

Data records and calculation results are filled in Table 4-2-1.

Table 4-2-1 Calibration of $Na_2S_2O_3$ standard solution and analysis results of water samples

	The experiment number	1	2	3
$KMnO_4$ standard solution calibration	$K_2Cr_2O_7$ volume/mL			
	Final burette reading/mL			
	Initial burette reading/mL			
	$Na_2S_2O_3$ consumed volume/mL			
	$1/6 K_2Cr_2O_7$ concentration/(mol/L)			
	$Na_2S_2O_3$ concentration/(mol/L)			

续表

	The experiment number	1	2	3
Water sample analysis	Water sample volume/mL			
	Initial burette reading/mL			
	Final burette reading/mL			
	$Na_2S_2O_3$ consumed volume/mL			
	Concentration of total residual chlorine(Cl_2, mg/L)			

6 Questions

(1) Why must there be a certain amount of residual chlorine in finished water of drinking water factories and distribution system?

(2) Why must titration take place in a weakly acidic solution with pH≈4?

Experiment 9 Determination of Cl⁻ in water (precipitation titration)

Chlorine ion (Cl^-) is a common inorganic anion in water and wastewater. Chlorine ions are present in almost all natural water, and their content range varies greatly. Chloride has important physiological functions and industrial use in human activities.

The determination of Cl^- is based on the method of precipitation titration, which must meet the basic requirements of titration analysis. The solubility of the precipitate formed by precipitation reaction must be small. Precipitation adsorption does not interfere with the determination of titration endpoint.

1 Experimental purpose

(1) Master the sources and content of chlorine ions in natural water and wastewater.

(2) Master the principle and method of precipitation titration.

2 Experimental principle

In neutral or weakly alkaline solution (pH=6.5-10.5), potassium chromite (K_2CrO_4) is used as an indicator to directly titrate Cl^- in water with $AgNO_3$ standard solution. Since the solubility of AgCl (8.72×10^{-8} mol/L) is less than that of Ag_2CrO_4 (3.94×10^{-7} mol/L), according to the principle of step precipitation, AgCl precipitation is first precipitated during titration and the precipitation reaction is as follows.

$$Ag^+ + Cl^- \longrightarrow AgCl \downarrow$$

(white precipitate)

When the stoichiometric point is reached, all Cl^- in the water has been titrated, and a little excessive Ag^+ and CrO_4^{2-} will generate Ag_2CrO_4 brick red precipitation, indicating that the end point of titration is reached. The precipitation titration reaction is as follows.

$$2Ag^+ + CrO_4^{2-} \longrightarrow Ag_2CrO_4 \downarrow$$
(brick red precipitation)

The content of Cl^- in the water sample is calculated according to the concentration and volume of $AgNO_3$ standard solution.

3 Experimental instruments and reagents

3.1 Instruments

Burette 50mL, titration table, conical flask 250mL, dropper, rubber suction bulb, glass rod, beaker 200mL.

3.2 Reagents

(1) Preparation of NaCl standard solution: A certain amount of NaCl is put into the crucible and heated for 40-50min at 500-600℃. After cooling, 8.2400g NaCl is weighed and dissolved in a small amount of distilled water. The solution is transferred into a 1000mL volumetric flask and diluted to scale with concentration of 0.0141mol/L. Transfer 10mL above solution and dilute it to 100mL with water. This solution contains 0.500mg chlorine ions (Cl^-) per mL.

(2) Preparation of $AgNO_3$ standard solution (0.1000mol/L): Weigh 2.395g $AgNO_3$, dissolve it in distilled water and dilute it to 1000mL. The solution is transferred to a brown reagent bottle and stored in the dark. The concentration of this solution is about 0.0141mol/L.

(3) Calibration of $AgNO_3$ standard solution: Three 25mL NaCl standard solution are transferred, and 25mL distilled water is taken as blank. Put them into 250mL conical flasks, respectively. Add 25mL distilled water and 1mL K_2CrO_4 indicator, and titrate the solution with $AgNO_3$ solution under constant shaking until the pale orange precipitation just appears and this is the end point. Record the volume of $AgNO_3$ solution. According to the concentration of NaCl standard solution and the volume of $AgNO_3$ solution, the standard concentration of $AgNO_3$ solution is calculated as follows:

$$c_1 = \frac{c_2 V_2}{V_1} \tag{4-2-3}$$

Where, c_1 ——concentration of $AgNO_3$ standard solution, mol/L;
V_1 ——titration volume of $AgNO_3$ standard solution, mL;
c_2 ——concentration of NaCl standard solution, mol/L;
V_2 ——volume of standard solution of NaCl, mL.

(4) Preparation of K_2CrO_4 indicator solution (5%): 5g K_2CrO_4 is dissolved in a small amount of water. $AgNO_3$ solution above mentioned is used to drop until the formation of red precipitation. Mix well. Stand for 12h, and the filtrate is filtered into a 100mL volumetric flask and diluted with distilled water to scale.

4 Experimental procedure

(1) If the pH values of the water samples are in the range of 6.5-10.5, they can be

titrated directly. For water samples beyond this range, phenolphthalein should be used as an indicator and the solution should be adjusted with 0.05mol/L H_2SO_4 solution or NaOH solution to pH≈8.0.

(2) If the content of organic matters or chromaticity in water samples is high, take 150mL water samples, put them into 250mL conical flasks, add 2mL aluminum hydroxide suspension respectively, shake and filtrate, and discard the initial filtrate of 20mL.

(3) If the water sample contains sulfide, sulfite or thiosulfate, adjust the water sample with sodium hydroxide solution to neutral or slightly alkaline. Then add 1mL H_2O_2 with a mass fraction of 30%, and mix well. After 1min, just heat the solution to 70-80℃ to remove excess H_2O_2.

(4) If the permanganate index in the water sample is greater than 15mg/L, add a small amount of $KMnO_4$, boil it, add a few drops of ethanol to remove the excess $KMnO_4$, then filter and sampling.

(5) Water sample analysis. Three portions of 50mL water samples and 50mL distilled water (blank experiment) are respectively put into the conical flasks. Add 1mL $K_2Cr_2O_4$ indicator and titrate the solution with $AgNO_3$ standard solution under violent shaking until the light orange color just appears, which is the end point. Record the volume of $AgNO_3$ standard solution.

5 Data processing

According to the following formula:

$$\text{mass concentration of chlorine ions}(Cl^-, mg/L) = \frac{(V_2 - V_0)c \times 35.453 \times 1000}{V_{\text{water}}}$$

(4-2-4)

Where, V_2 ——volume of $AgNO_3$ standard solution consumed by water samples, mL;

c ——concentration of $AgNO_3$ standard solution, mol/L;

V_0 ——volume of $AgNO_3$ standard solution consumed by distilled water, mL;

V_{water} ——volume of water sample, mL;

35.453——molar mass of Cl^-, g/mol.

The experimental results and calculation results are filled in Table 4-2-2.

Table 4-2-2 Calibration of $AgNO_3$ standard solution and analysis results of water samples

	The experiment number	1	2	3
Solution calibration	Volume of NaCl standard solution/mL			
	Titration end point reading/mL			
	Titration start point reading/mL			
	Titration volume of $AgNO_3$ standard solution/mL			
	Concentration of NaCl standard solution/(mol·L^{-1})			
	Concentration of $AgNO_3$ standard solution/(mol·L^{-1})			

续表

	The experiment number	1	2	3
Determination of water	Volume of the water sample/mL			
	Titration end point reading/mL			
	Titration start point reading/mL			
	Titration volume of AgNO$_3$ standard solution/mL			
	Concentration of chlorine ions(Cl$^-$)/(mg·L^{-1})			

6 Questions

(1) Why is the determination of Cl$^-$ in water is in neutral or weakly alkaline solution by Mohr method?

(2) When $K_2Cr_2O_4$ is used as indicator, what is the effect of too high or too low concentration of indicator on the determination?

(3) Why is it necessary to shake violently when titrating Cl$^-$ with AgNO$_3$ standard solution?

Experiment 10 Fenton reagent catalyzed oxidation of dye wastewater

1 Experimental purpose

(1) Understand the properties of Fenton reagent.

(2) Understand the degradation mechanism of organic pollutants by Fenton reagent.

(3) To grasp the influence of various factors in Fenton reaction on the decolorization rate of wastewater.

2 Experimental principle

The oxidation mechanism of Fenton reagent can be expressed in the following chemical reaction equation:

$$Fe^{2+} + H_2O_2 \longrightarrow Fe^{3+} + OH^- + OH\cdot$$

The formation of OH· gives Fenton reagent a strong oxidation capacity. Researches have shown that in the solution of pH=4, its oxidation ability is second only to fluorine gas in aqueous solution. Therefore, persistent organic pollutants, especially aromatic compounds and some heterocyclic compounds, can be oxidized and decomposed by Fenton reagent. In this experiment, fenton reagent is used to treat methyl-orange simulated dye wastewater. A certain concentration of methyl-orange simulated wastewater is prepared. During the experiment, the wastewater is taken into a beaker (or conical flask), a certain amount of ferrous sulfate is added, and a constant temperature magnetic stirrer is turned on to make it fully mixed and dissolved. After being dissolved, add quantitative

H_2O_2 rapidly, blend, react to the given time, terminate reaction with NaOH solution, adjust pH value to 8-9, stand for the appropriate time, take the supernatant and measure the absorbance at the maximum absorption wavelength of 465nm. The chroma removal rate = (absorbance difference at maximum absorption wavelength before and after reaction/absorbance before reaction)× 100%.

3 Experimental instruments and reagents

3.1 Instruments

(1) Acidity meter or pH test paper.

(2) 722 visible light spectrophotometer.

(3) Tray balance, analytical balance, fifteen 250mL conical flasks, five 2mL pipettes, four 5mL pipettes, four 100mL measuring cylinders.

3.2 Reagents

(1) Methyl-orange.

(2) $FeSO_4 \cdot 7H_2O$, H_2O_2 (30%), H_2SO_4, and NaOH (all are analytically pure).

4 Experimental procedure

(1) Prepare 200mg/L methyl-orange simulated wastewater. During the experiment, 200mL 200mg/L methyl orange simulated wastewater is placed in a beaker (or conical flask).

(2) Determine the appropriate amount of ferrous sulfate. Specific practice is as follows. The concentration of methyl-orange simulated wastewater is 200mg/L, the dosage of H_2O_2 (30%) is 1mL/L, and the pH of the water sample is 4.0-5.0. When the temperature of the water sample is room temperature, different amounts of $FeSO_4 \cdot 7H_2O$ (20mg/L, 60mg/L, 100mg/L, 200mg/L, 300mg/L, respectively) are added for decolorization experiment, and the reaction time is 90min. The optimal dosage of $FeSO_4 \cdot 7H_2O$ is determined by this experiment.

(3) Determine the appropriate amount of H_2O_2 (30%). Specific practice is as follows. The concentration of methyl-orange simulated wastewater is 200mg/L, the dosage of $FeSO_4 \cdot 7H_2O$ is the optimal dosage determined in (2), the pH value of the water sample is 4.0-5.0, the temperature of the water sample is room temperature, different amounts of H_2O_2 (30%) (0.1mL/L, 0.2mL/L, 0.4mL/L, 0.6mL/L, 0.8mL/L) are added for decolorization experiment, and the reaction time is 90min. Through this experiment, the optimal dosage of H_2O_2 (30%) is determined.

(4) Determine the effect of pH on the degradation effect. The concentration of methyl-orange simulated wastewater is 200mg/L, the amount of $FeSO_4 \cdot 7H_2O$ is the optimal amount determined in (2), and the amount of H_2O_2 (30%) is the optimal amount determined in (3). The effect of pH (pH is 1, 2, 3, 4, 5, 6, respectively) on the degradation effect of methyl-orange simulated wastewater is investigated to determine the optimal pH.

(5) Determine the effect of reaction time on the degradation effect. The Concentration of methyl-orange simulated wastewater is 200mg/L, the pH value of the water sample is

4.0, the dosage of $FeSO_4 \cdot 7H_2O$ is the optimal dosage determined from the above experiments, the dosage of H_2O_2 (30%) is the optimal dosage determined from the above experiments, and the effect of the reaction time (sampling times of 10min, 20min, 40min, 60min, and 120min) on the degradation of the simulated wastewater of methyl orange is investigated under the optimal pH conditions.

5 Data processing

The chroma removal rate is calculated as shown below:

$$\text{chroma removal rate} = \frac{A_0 - A}{A_0} \quad (4\text{-}2\text{-}5)$$

Where, A_0——the absorbance of the sample before treatment;
A——the absorbance of the sample after treatment.

6 Questions

The effect of various factors in the Fenton reaction on the decolorization rate of wastewater is investigated.

7 Additional content

Azo dyes are the largest part of industrial dyes. Methyl-orange, as a representative acid azo dye, is widely used at present. It is used as reagent and indicator chemically, and used for dyeing cotton, hemp, paper and leather in industry. Methyl-orange has a stable structure and is difficult to volatilize and biodegrade. The molecular formula of methyl-orange is $C_{14}H_{14}N_3NaO_3S$ with a molecular weight of 327.33. Its chemical structure is shown in Fig. 4-2-1. Methyl-orange dye is weakly acidic with a maximum absorption at the wavelength of 464nm, as shown in Fig. 4-2-2.

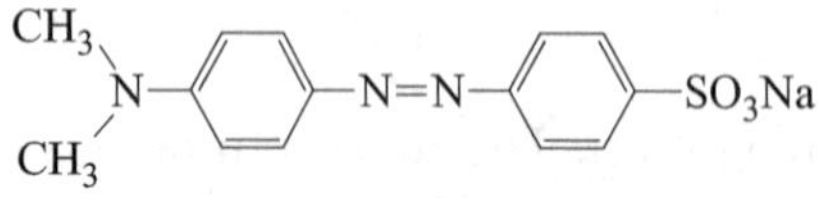

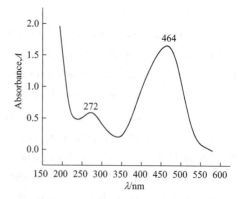

Fig. 4-2-1 Chemical structure of methyl orange

Fig. 4-2-2 Ultraviolet-visible absorption wavelength spectrogram of methyl-orange

Experiment 11 Existing forms of copper in natural water

1 Experimental purpose

(1) Distinguish several simple combined states of copper in lake water.

(2) Learn the general experimental technique for determination of metal combined state in water by anodic dissolution voltammetry.

(3) Be familiar with and master the use of XJP-821 (B) new polarograph.

2 Experimental principle

The existing forms of heavy metals in natural water can be divided into particle state and dissolved state according to their physical state. The former includes various chemical states of adsorption and complexation on suspended particles. The latter is divided into stable state and unstable state according to their activity in water. The unstable state mainly includes free metal ions, weakly bound organic and inorganic complexed adsorbed metals. The metals in this state are electrically active and reacts on the electrode. The stable state mainly includes the strongly bound organic and inorganic complexed adsorbed metal. They don't react on the electrodes. But after exposure to ultraviolet light, the organic bonding state, i. e., stable state A, becomes unstable. The part that is not decomposed by ultraviolet light, namely stable state B, will also become unstable after being digested by nitric acid-perchloric acid. The unstable state is electrically active and can be enriched by microelectrodes, which can be determined by dissolution voltammetry.

With the proper base solution and applied voltage, the unstable copper can be reduced to copper and deposited on the working electrode. That is, the unstable copper is electrically active and can be enriched by electrolysis. The stable copper does not react with the electrode and cannot be enriched. This is an important basis for electrochemistry to distinguish the stable and unstable states of heavy metals in natural water. With ultraviolet irradiation and strong acid digestion, the stable state can be further divided into A and B states. For copper film enriched on the working electrode, when the potential reaches the dissolution potential of copper in the scanning process from negative potential to positive potential, the copper is rapidly oxidized into copper ions, which are immersed in the solution and form a dissolution current peak at the same time. When other conditions remain unchanged, the content of copper in the measured liquid can be determined according to the height of the dissolution current peak.

This experiment is to determine the binding state of copper in lake water slightly polluted by copper. The lake water is filtered with 0.45μm membrane and the copper content is determined with XJP-821 (B) new polarograph and ATA-1A rotating disc electrode of polarographic technology respectively in: ①untreated filtrate; ②filtrate treated with ultraviolet light; ③filtrate digested with nitric acid and perchloric acid; ④suspension digested with nitric acid and perchloric acid.

Obviously, only the unstable copper is measured in the untreated filtrate. Unstable state copper and the stable state A copper are measured in the filtrate treated with ultraviolet light. After digestion with nitric acid and perchloric acid, they includes unstable state copper, stable state A copper and stable state B copper. Therefore, the content of copper in unstable state, stable state A, stable state B, and granular state can be determined and calculated respectively.

3 Experimental instruments and reagent

XJP-821 (B) new polarography, ATA-1A rotating disc electrode, UV lamp, electric heating plate (800W), microporous membrane filter, vacuum pump system, nitrogen cylinder, pipettes (0.25mL, 2mL, 5mL, 50mL, 100mL), 100mL volumetric cylinder, concentrated $HClO_4$ (GR), concentrated HNO_3 (GR), 30% H_2O_2 (GR), copper standard liquid (1.00mg/L), base solution (3mol/L NH_4Cl-NH_4Ac-$NH_3 \cdot H_2O$).

4 Experimental procedure

(1) Water sample collection and treatment: Take 1L lake water with a plastic bucket and filter large particles with glass fiber. Then use a 100mL pipette to take 200mL water sample and put it into a microporous membrane filter for filtration. Discard the first 10mL filtrate and store the rest filtrate in the refrigerator for later use. Remove carefully microporous membrane and suspended material. Put them into a 150mL beaker and set aside.

(2) Light treatment of filtrate: Transfer 50mL filtrate into the ultraviolet lamp, add 10 drops of 30% H_2O_2, and carefully put in the stirring magnetons. Connect cooling water and power supply, and light filtrate for 2hours during stirring. Pour the filtrate into a 100mL beaker and set aside.

(3) Digestion of filtrate: Transfer 50mL filtrate into a 150mL beaker, add 2mL concentrated HNO_3, and boil it on the electric heating plate. When the test solution is nearly dry, add 2mL concentrated $HClO_4$, and continue to heat until the white smoke is exhausted and the contents are nearly dry. Remove the beaker, cool it, add 10mL distilled water, and boil it on the electric heating plate for 1minute. Cool and transfer it into a 100mL volumetric flask, add distilled water to constant volume, and set aside.

(4) Treatment of filter membrane: Add 10mL distilled water and 2mL concentrated nitric acid into the beaker containing microporous membrane and suspended matters, and then continue the treatment process as filtrate digestion.

(5) Determination: Transfer 50.0mL solution into an electrolytic cell cup and add 5.0mL 3mol/L NH_4Cl-NH_4Ac-$NH_3 \cdot H_2O$ base solution, insert ATA-1A rotating disc electrode into the electrolytic cell, adjust the electrode speed, and pass through the high purity nitrogen to remove oxygen for 5min. At the same time, seal the electrolytic cell system, press the electrode switch, and trigger the enrichment key. After 2min, the instrument will stop to automatically scan for 30 s to obtain a dissolution peak. Measure the peak height as

h, and repeat the above operation for 3-4 times (note that electrochemical cleaning of the electrode must be conducted before each measurement). Then, 0.25mL 1.00mg/L copper standard solution is added into the electrolytic cell cup, and the determination is repeated for 3-4 times under the same conditions, The peak height as H is recorded. Anodic dissolution voltammetry determination conditions is shown in Table 4-2-3. Filtrate illumination samples, filtrate digestion samples and filter membrane samples are determined as above.

Table 4-2-3 Anodic dissolution voltammetry determination conditions

Upper limit potential/V	−1.0	Enrichment time/min	2
Initial potential/V	−0.9	Electrode speed/(r/min)	1000
Lower limit potential/V	0	X-axis range/(mV/cm)	100
Scanning speed/(mV/s)	100	Y-axis range/(mV/cm)	50

5 Data processing

(1) Calculate the copper content of samples treated in different ways according to the formula (4-2-6) and (4-2-7), and fill the results in Table 4-2-4.

$$c'_x = \frac{nc_0 h}{H(m+n) - hm} \tag{4-2-6}$$

$$c_x = \frac{mc'_x}{v} \times 1000 \tag{4-2-7}$$

Where, c'_x——sum of the concentration of copper in test solution and base solution, μg/mL;

h——peak height of test solution, mm;

H——peak height of test solution after adding standard solution, mm;

v——volume of test solution, mL;

m——sum of the volume of test solution and base solution, mL;

n——volume of added copper standard solution, mL;

c_0——mass concentration of added copper standard solution, mg/L;

c_x——mass concentration of copper in test solution, μg/L.

Table 4-2-4 Determination of copper content treated in different ways

Treatment ways	Deal directly	Light processing	Digestion process	Membrane treatment
Peak height of test solution, h/mm				
Peak height of test solution after adding standard solution, H/mm				
Mass concentration of copper/(μg/L)				

(2) According to the mass concentration of copper obtained by various treatment methods, the mass concentration of copper in the lake in the granular state, unstable state, stable state A and stable state B are calculated, and the results are filled in Table 4-2-5.

Table 4-2-5 The mass concentration of copper in different states in lake

Form of a state	Particle state	Dissolved state		
		Unstable state	Stable state A	Stable state B
Mass concentration of copper/(μg/L)				

6 Questions

(1) Why can polarography be used to determine metal forms in water?

(2) Briefly describe the reason for the high sensitivity of anode dissolution voltammetry.

(3) In order to ensure the accuracy of measurement, what key issues should be paid attention to?

7 Notes

(1) Since the sources of lake water are different and the content of copper is different, the volume of removed sample and the volume of added standard solution can be adjusted appropriately according to the specific situation.

(2) When the test solution is digested with HNO_3 and $HClO_4$, it should be heated until the white smoke is exhausted after digestion, but do not steam it dry. If there is too much acid left in the solution, the acidity of the test solution is so high that it affect the determination. If evaporate it dry, local temperature is so high that trace copper may evaporate.

(3) After each dissolution measurement, electrochemical cleaning should be carried out on the electrode, and the next measurement should be made after scanning and confirming that there is no copper.

(4) In order to obtain the dissolution current peak with good reproducibility for each sample, it is necessary to ensure that the enrichment time, static time, electrode cleaning time, etc. are strictly consistent.

Experiment 12　Determination of biochemical oxygen demand (BOD_5) in wastewater

1　Experimental purpose

(1) Master the collection and treatment of water samples.

(2) Master measurement principle and operation of BOD_5.

2　Experimental principle

Take two domestic sewage samples. The one is to determine the dissolved oxygen at that time. The other sample is cultured at (20 ± 1)℃ for 5 days before the determination of dissolved oxygen. The difference between the two is BOD_5.

The determination principle of dissolved oxygen: Add manganese sulfate and basic potassium iodide in the water sample. Bivalent manganese first generate the white $Mn(OH)_2$ precipitation, which soon oxidized by dissolved oxygen in the water to trivalent or quadruvalent manganese, so that the dissolved oxygen is fixed. Under acidic conditions, high-value manganese can oxidize I^- to I_2, and then I_2 is titrated with sodium thiosulfate standard solution to find the dissolved oxygen content in water.

3 Experimental instruments and reagents

3.1 Instruments

(1) Thermostatic incubator.

(2) The dissolved oxygen bottle (200-300mL) is provided with a ground glass plug and a bell mouth with a closed water supply.

3.2 Reagents

(1) Manganese sulfate solution: Weigh 480g manganese sulfate ($MnSO_4 \cdot H_2O$) and dissolve it in water, and dilute it to 1000mL with water. This solution is added to acidified potassium iodide solution, and shall not produce blue color with starch.

(2) Alkaline potassium iodide solution: Weigh 500g sodium hydroxide and dissolve it in 300-400mL water. Weigh 150g potassium iodide and dissolve it in 200mL water. After the sodium hydroxide solution is cooled down, the two solutions will be mixed well and diluted with water to 1000mL. If there is a precipitate, place the solution overnight, pour out the upper layer of the clear solution, store it in a brown bottle, corke the bottle tightly with a rubber stopper, and store it away from light. After acidification, this solution should not show blue color with starch.

(3) Sodium thiosulfate solution: Weigh 2.5g sodium thiosulfate ($Na_2S_2O_3 \cdot 5H_2O$) and dissolve it in boiled but cooled water, add 0.2g sodium carbonate, dilute it with water to 1000mL, and store it in a brown bottle. Before use, calibrate it with potassium dichromate standard solution.

(4) Concentrated sulfuric acid ($\rho = 1.84 g/mL$).

(5) 0.5% starch solution: Weigh 0.5g soluble starch, make a paste with a small amount of water, and then dilute it with freshly boiled water to 100mL. After cooling, add 0.1g salicylic acid and 0.4g zinc chloride for preservation.

4 Experimental procedure

(1) Sample collection: Prepare six dissolved oxygen bottles, transfer the water samples to the dissolved oxygen bottles, and allow the water samples to overflow the mouth of the bottles for a few seconds. Three of the bottles are used to determine the dissolved oxygen by fixing the dissolved oxygen. The other three bottles are cultivated in a constant temperature incubator for 5 days before the dissolved oxygen are measured.

(2) Dissolved oxygen fixation: Insert the pipette under the liquid level of the dissolved oxygen bottle and add 1mL manganese sulfate solution and 2mL alkaline potassium iodide solution. Plug the bottle, mix it upside down several times, and let it stand.

(3) Determination of dissolved oxygen: Uncork the bottle, immediately insert the pipette into the liquid, and add 2.0mL concentrated sulfuric acid. Plug the bottle, and mix it upside down and shake well until all the precipitate is dissolved. Put it in the dark and let it stand for 5 minutes.

Transfer 100.00mL above solution into a 250mL conical flask, titrate the solution with sodium thiosulfate standard solution until the solution is pale yellow, add 1mL starch solution, continue to titrate until the blue color just fades, and record the volume of sodium thiosulfate solution.

5 Data processing

$$\text{mass concentration of dissolved oxygen}\left(O_2, \frac{\text{mg}}{\text{L}}\right) = \frac{1}{4} \times \frac{cVM_{O_2}}{V_0} \times 10^3 \quad (4\text{-}2\text{-}8)$$

Where, c ——concentration of sodium thiosulfate standard solution, mol/L;

V ——volume of sodium thiosulfate standard solution comsumed by titration, mL;

V_0 ——volume of the water sample taken when titrating, mL;

M_{O_2} ——molar mass of O_2, g/mol.

6 Notes

(1) If the pH value of water sample exceeds the range of 6.5-7.5, dilute solution of hydrochloric acid or sodium hydroxide can be used to adjust it to 7, but the dosage should not exceed 0.5% of the volume of water sample.

(2) Water samples collected from water with relatively low water temperature are found to contain supersaturated dissolved oxygen. The water samples should be rapidly heated to about 20℃ and shaken sufficiently to drive out saturated dissolved oxygen.

If water samples are obtained from a water bath or sewage outlet with higher temperature, they should be cooled rapidly to about 20℃ and shaken sufficiently to be in balance with the oxygen partial pressure in the air.

Experiment 13　Determination of chromium in industrial wastewater (diphenylcarbazide spectrophotometry)

1 Experimental purpose

(1) Master the principle and method of determination of hexavalent chromium and total chromium in water by diphenylcarbazide spectrophotometry.

(2) Learn how to get linear regression equation with Microsoft Office Excel.

2 Experimental principle

In acidic solution, hexavalent chromium ions react with diphenylcarbazide to form a purple compound with a maximum absorption at the wavelength of 540nm. The relation between absorbance and concentration conforms to Beer's Law. If the total chromium is determined, the trivalent chromium in the water sample should be oxidized to hexavalent chromium with potassium permanganate before the determination.

3 Experimental instruments and reagents

3.1 Instruments

Spectrophotometer.

3.2 Reagents

(1) Acetone.

(2) (1+1) sulfuric acid.

(3) (1+1) phosphoric acid.

(4) 2g/L sodium hydroxide solution.

(5) Zinc hydroxide coprecipitation agent: Zinc sulfate ($ZnSO_4 \cdot 7H_2O$) is weighed of 8g and dissolved in 100mL water. Weigh 2.4g sodium hydroxide and dissolve it in 120mL water. Mix the two solutions together.

(6) 40g/L potassium permanganate solution.

(7) Chromium standard storage solution: Weigh 0.2829g potassium dichromate (guarantee reagent), dried at 120℃ for 2hours, dissolve it in water, transfer it into a 1000mL volumetric flask, and dilute it with water to marked line.

(8) Chromium standard solution: Transfer 5.00mL chromium standard stock solution into a 500mL volumetric flask, dilute it with water to the standard line, and shake well. Each milliliter of standard solution contains 1.000μg hexavalent chromium. Prepare the solution on the day of use.

(9) 200g/L urea solution.

(10) 20g/L sodium nitrite solution.

(11) Diphenylcarbazide solution: 0.2g diphenylcarbazide dihydrazide (DPC, $C_{13}H_{14}N_4O$) is weighed and dissolved in 50mL acetone, and diluted with water to 100mL. Shake it well, and store it in a brown bottle in the refrigerator. It can't be used when the color gets darker.

(12) Nitric acid.

(13) Sulfuric acid ($\rho = 1.84$g/mL).

(14) Trichloromethane.

(15) (1+1) ammonia.

(16) 50g/L cupferron: 5g Cu-fe reagent [$C_6H_5N(NO)ONH_4$] is weighed and dissolved in ice water, then diluted to 100mL. Prepare the solution at the time of use.

4 Experimental procedure

4.1 Determination of hexavalent chromium

4.1.1 Water sample pretreatment

(1) For clean surface water that is free of suspended solids and low in color, the measurement is made directly.

(2) If the water sample is colored but not dark, color correction can be carried out. Take another sample, add various reagents other than chromogenic agent, replace chromogenic agent with 2mL acetone, and use this solution as the reference solution to determine the absorbance of the sample solution.

(3) For turbid and dark water samples, zinc hydroxide coprecipitant agent should be added and filtered.

(4) In the presence of hypochlorite and other oxidizing substances in the water samples, urea and sodium nitrite can be added to eliminate interference.

(5) In water containing reducing substances, such as ferrous iron, sulfite, and sulfite, Cr(VI) can be reduced to Cr(III). At this time, adjust the pH value of the water sample to 8, add the color developing agent solution, place it for five minutes, then acidify the color development, and draw the standard curve in the same way.

4.1.2 Standard curve drawing

Take nine 50mL colorimetric tubes, successively add 0.00, 0.20, 0.50, 1.00, 2.00, 4.00, 6.00, 8.00, 10.00mL chromium standard solution, and dilute them with water to the marked line. Add 0.5mL (1+1) sulfuric acid and 0.5mL (1+1) phosphoric acid in the solution, and shake well. Then add 2mL color developing agent solution and shake well. 5-10min later, the absorbance is measured and blank correction is performed with water as the reference in a 1cm or 3cm covette at the wavelength of 540nm. Take absorbance as the ordinate and the corresponding content of hexavalent chromium as the abscissa to draw the standard curve with Microsoft Office Excel and get the linear regression equation.

4.1.3 Determination of water sample

Take an appropriate amount of colorless and transparent or pretreated sampling water (containing less than $50\mu g$ chromium) in a 50mL colorimetric tube, dilute it with water to the marked line, and measure it in the same way as the standard solution. After blank correction, the mass of Cr(VI) can be found on the standard curve according to the measured absorbance.

4.1.4 Calculation

The mass concentration of chromium ρ (mg/L) in water samples is calculated according to equation (4-2-9):

$$\rho\left(Cr, \frac{mg}{L}\right) = \frac{m}{V} \tag{4-2-9}$$

Where, m——mass of Cr(VI) found on the standard curve, μg;

V——volume of water sample, mL.

4.2 Determination of total chromium

4.2.1 Water sample pretreatment

General clean surface water can be directly determined after potassium permanganate oxidation.

Water samples containing a lot of organic matters need to be digested. Take 50mL or an appropriate amount of water (containing less than $50\mu g$ chromium), put it into a 150mL beaker, add 5mL nitric acid and 3mL sulfuric acid, and heat and evaporate until white smoke. If the solution is still colored, add 5mL nitric acid and repeat the above operations until the solution is clear and cool. Dilute it with water to 10mL, neutralize it with ammonia (1+1) to pH=1-2, transfer the solution into a 50mL volumetric flask, dilute it with water to the marked line, shake well, and set aside.

If the content of molybdenum, vanadium, iron, copper and so on in the water sample is large, cupferron-trichloromethane extraction is used to remove them, and then the digestion is processed.

4.2.2 Oxidation of trivalent chromium by potassium permanganate

Take 50.0mL or an appropriate amount of (chromium content less than $50\mu g$) clean water or pre-treated water (less than 50.0mL, water will be added to 50.0mL) in a 150mL conical flask, adjust it with (1+1) ammonia and sulfuric acid solution to neutral, add a few glass beads, add 0.5mL (1+1) sulfuric acid and 0.5mL (1+1) phosphoric acid, and shake well. Add two drops of 40g/L potassium permanganate solution. If the purple fades, continue to add potassium permanganate solution until it remains purplish red. Heat and boil until solution is about 20mL. After cooling, add 1mL 200g/L urea solution and shake well. Add 20g/L sodium nitrite solution with dropper and shake well with each drop until the purple color just disappears. Stop for a moment, wait for the bubble in the solution to escape, transfer it to a 50mL colorimetric tube, and dilute it to the marked line for measurement.

The other steps are the same as the determination of hexavalent chromium.

5 Data processing

(1) Filling the experimental results in Table 4-2-6.

Table 4-2-6 Experimental data record

	Standard curve drawing									Determination of water
Water sample volume/mL	0	0.2	0.5	1.0	2.0	4.0	6.0	8.0	10.0	
Absorbance										
Cr mass/μg										

(2) Draw standard curve.

(3) Calculate the content of Cr in water samples.

6 Notes

(1) Glassware used for the determination of chromium shall not be washed with potassium dichromate lotion.

(2) For the chromogenic reaction of Cr(Ⅵ) and chromogenic agent, generally control the acidity in the range of 0.05-0.3mol/L (1/2 H_2SO_4). When it's 0.2mol/L, color development is best. Water samples should be adjusted to neutral before color development. Color development temperature and starding time have an effect on color development. The color can be stable in 5-15min at 15℃.

(3) For the determination of clean surface water samples, the chromogenic agent can be prepared in the following ways: dissolve 0.2g diphenylcarbazide in 100mL 95% ethanol, and add 400mL (1+9) sulfuric acid while stirring. The solution will keep in the refrigerator for one month. With this chromogenic agent, 2.5mL can be directly added in the color developing process, without adding acid. But after adding chromogenic agent, it needs to be immediately shaked, so as Cr(Ⅵ) not to be reduced by acetic acid.

Experiment 14 Coagulation experiment

The colloidal particles dispersed in water have electric charges. Under the action of Brownian motion and its surface hydration film, they are in a stable and dispersive state for a long time and cannot be removed by natural precipitation method. When the coagulant is added to the water, the dispersed particles can be combined with each other and become aggregated and separated from the water.

Due to the great difference of raw water, coagulation effect is not the same. The coagulation effect of coagulant depends not only on the amount of coagulant added, but also on the pH value of water and the gradient of flow velocity.

1 Experimental purpose

(1) Observe coagulation phenomenon and process, understand the mechanism of water purification and important factors affecting coagulation.

(2) Master the basic method to obtain the optimal coagulation conditions (dosage and pH) of a water sample.

2 Experimental principle

Suspended matters and colloidal substances with small particle size in water are stable in turbidity state due to Brownian motion of particles, electrostatic repulsion between colloidal particles, and surface action of colloid. The treatment objects of chemical coagulation are mainly the tiny suspended matters and colloids in wastewater. According to the characteristics of colloid, the stability of colloid is usually destroyed by adding electrolyte, colloidal particles with different charges, or macromolecules in the process of wastewater treatment. Then the wastewater is purified by precipitation separation. There are four main explanations for the mechanism of chemical coagulation.

2.1 Mechanism of compression of double electric layers

When two colloidal particles are close to each other and the double electric layers overlap, electrostatic repulsion is generated. The electrostatic repulsion between the added anti-ions and the original anti-ions in the diffusion layer squeezes part of the anti-ions into the adsorption layer, thus it reduces the thickness of the diffusion layer. As the diffusion layer thins, the particles collide at less distance and attract each other more. The net force of repulsive force and attractive force between particles changes from repulsive force to attractive force, and particles can condense with each other.

2.2 Mechanism of adsorption electroneutralization

The different charge of colloidal particles attract each other to achieve electrical neutralization and condensation. Big colloidal particles adsorb many small colloidal particles or ions with different charges, ζ potential decreases, and attractive force makes colloidal particles with different charges close to each other and coagulate.

2.3 Mechanism of adsorption bridging

Adsorption bridging refers to the phenomenon that the chain polymer is connected with colloidal particles and fine suspended matters through the active part under the action of electrostatic attraction, van der Waals force and hydrogen bonding force.

2.4 Sediment trapping mechanism

When high-value metal salts such as aluminum salt or iron salt are used as coagulants, a large amount of metal hydroxide precipitation will generate with large dosage, and the colloidal particles in the water can be netted and scrolled. These deposits are the core of the colloidal particles in water. It's basically mechanical.

After adding coagulant to the water, the results are as follows: ① It can reduce the repulsion energy peak between particles and the ζ potential of colloidal particles, and destabilize the colloidal particles. ② At the same time polymer type polymer coagulant adsorption bridge action can also occur. ③ It can achieve the aggregation of particles by net effect.

The process of removing or reducing the stabilizing factors of colloidal particles is called destabilization. Under certain hydraulic conditions, the colloidal particles after destabilization can form larger flocs, commonly known as alum flocs. Larger diameter and more dense alum flocs are easy to sink. The process of adding coagulants to form larger alum flocs is called coagulation. In the process of coagulation, the above phenomenon often exist at the same time, only in a certain situation to some certain phenomenon.

3 Experimental instruments and reagents

(1) Intelligent coagulation test blender.
(2) PHS-2 type acidity meter.
(3) HACH 2100N turbidimeter.

(4) Beakers (200mL, seven).

(5) Pipettes (1mL, 2mL, 5mL, 10mL each).

(6) One rubber suction bulb is used for medicinal transfer with pipettes.

(7) One 1000mL measuring cylinder.

(8) Coagulants: Aluminum sulfate [$Al_2(SO_4)_3$], polyferric sulfate (PFS), polyaluminum chloride (PAC), polyferric aluminum sulfate (PAFS), polyacrylamide (PAM), etc. The concentration is 1% or 10g/L.

(9) Hydrochloric acid and sodium hydroxide (concentration is 10%).

(10) Raw water for experiment (take 20L river water or mix clay and tap water for 20L water sample, settle for 6h, and the supernatant is raw water for experiment).

4 Experimental procedure

Coagulation experiment is divided into three parts: the best dosage, the best pH and the best flow velocity gradient. In the experiment of the best dosage, a change mode of stirring speed and pH are selected first, and the best dosage is calculated. Then find the best coagulation pH according to the optimal dosage of dosage.

4.1 Experimental procedure of optimal dosage

(1) Determine the characteristics of raw water, that is, determine turbidity, pH and temperature of raw water samples.

(2) Determine the minimum coagulation dose for the formation of alum flocs. The method is to stir 800mL raw water in a beaker at a low speed (50r/min) and increase the amount of coagulants by 1mL every 1min until alum flocs appear. The coagulation dose at this time is the minimum dosage for the formation of alum flocs.

(3) Six 1000mL beakers are used to place 800mL raw water on the platform of the blenders for coagulation test.

(4) Determine the dosage of coagulants during the experiment. According to step (2), the minimum dosage of coagulant for forming alum flocs is obtained. 1/4 was taken as the dosage of coagulants for No. 1 beaker, and 1/2, 3/4, 1/1, 3/2 and 2/1 times are taken as the dosage of coagulants for No. 2-6 beakers, respectively. When dosing, the coagulants should be added into the No. 1-6 dosing tubes of the instruments, so that the dosing can be guaranteed at the same time.

(5) Start the blender and stir rapidly for 30s at a speed of about 300r/min. Stir at medium speed for 6min at about 100r/min. Stir at low speed for 6min at about 50r/min. If doing coagulation experiments with sewage, stirring speed can be slowed down, because sewage colloidal particles are relatively fragile.

(6) Shut down the blender, lift up the stirring paddle, let the precipitation stand for 10 minutes, take out the supernatant and put it into the beaker, and immediately measure the turbidity with the turbidity meter (three times for each cup of water sample), which is recorded in Table 4-2-7.

Chapter 4 Environmental chemistry experiments

4.2 Experiment procedure of optimal pH

(1) Six 1000mL beakers are used to place 800mL raw water on the platform of the blenders for coagulation test.

(2) Adjust the pH of raw water and add 1.5mL, 1.0mL and 0.5mL 10% hydrochloric acid into the No.1, No.2 and No.3 beakers with water samples successively with pipettes. Add 0.5mL and 1.0mL 10% sodium hydroxide into the No.5 and No.6 beakers with water samples respectively.

(3) Start the blender and stir quickly for 30s at a speed of about 300r/min. The pH of each water sample is determined with an acidity meter, which is recorded in Table 4-2-8.

(4) The dosing tube of the instrument is used to add the same dose of coagulant into each beaker (the optimal dosage is obtained from experiment 4.1).

(5) Start the blender and stir rapidly for 30s at a speed of about 300r/min. Stir at medium speed for 6 min at about 100r/min. Stir at low speed for 6min at about 50r/min. If doing coagulation experiments with sewage, stirring speed can be slowed down, because sewage colloidal particles are relatively fragile.

(6) Shut down the blender, lift up the stirring paddle, let the precipitation stand for 10 minutes, take out the supernatant and put it into the beaker. The turbidity is measured immediately with the turbidity meter (three times for each cup of water sample), and recorded in Table 4-2-8.

5 Data processing

Coagulants _____   Coagulant concentration _____
Turbidity of raw water _____   Raw water pH _____
Raw water temperature _____
Minimum coagulant quantity (ml) _____   Equivalent to (mg/L) _____

5.1 Sorting out experimental results of optimal dosage

(1) The characteristics of raw water, the dosing condition of coagulants, and the residual turbidity after precipitation are recorded in Table 4-2-7.

Table 4-2-7 Optimal dosage of coagulants

Sample number		1	2	3	4	5	6
Dosage/mL							
Time of alum flocs appear/min							
Precipitation of alum flocs(fast/slow)							
Residual turbidity/NTU	1						
	2						
	3						

(2) The relation curve between residual turbidity and dosage is drawn with residual turbidity as the ordinate and dosage as the abscissa, from which the optimal dosage not greater

than a certain residual turbidity could be obtained.

5.2 Sorting out the experimental results of optimal pH

(1) The characteristics of raw water, the dosage of coagulants, the dosage of acid and alkali, and the residual turbidity after precipitation are recorded in Table 4-2-8.

Table 4-2-8   Optimal pH (dosage: _____ mL)

Sample number		1	2	3	4	5	6
HCl volume/mL							
NaOH volume/mL							
Water pH							
Residual turbidity/NTU	1						
	2						
	3						

(2) The relation curve between residual turbidity and pH is drawn with residual turbidity as the ordinate and water sample pH as the abscisate, and the optimal coagulant pH and its applicable range are obtained from the curve.

6 Questions

(1) According to the experimental results and the observed phenomena, the main factors affecting coagulation are briefly described.

(2) Why is the coagulation effect not necessarily good when the maximum dosage is applied?

(3) According to the experimental curve of optimal dosage, the relationship between turbidity of precipitated water and dosage of coagulants is analyzed.

7 Notes

(1) In the optimal dosage, the optimal pH experiments, it is required to add the coagulants at the same time into each beaker to avoid the difference of reaction time and coagulation effect due to long time interval.

(2) When determining the turbidity of water and pumping supernatant, do not disturb the bottom sediment. Meanwhile, the suction time intervals between beakers should be minimized.

Experiment 15   Determination of total organic carbon in water by nondispersive infrared absorption method

Total organic carbon refers to the total amount of carbon contained in dissolved and suspended organic matters in water, and it is usually denoted by TOC. There are many kinds of organic matters in water, which contain carbon, hydrogen, nitrogen, sulfur and other elements. They still cannot all be separated and identified. TOC is a comprehensive index for rapid detection, which indicates the total amount of organic matters in water in terms of the amount of carbon. However, because it cannot reflect the types and composition of organic matters in water, it cannot reflect the different pollution consequences caused by the same amount of TOC. Because TOC is determined by combustion method, it can oxidize all organic matters. It represents the total amount of organic matters more directly than BOD_5 or COD. It is usually used as an important basis to evaluate the degree of organic pollution in water. TOC analysis has become the main projects of water treatment and quality control in many countries around the world.

TOC is the amount of carbon that represents the total amount of organic matters in water, and the result is expressed with the mass concentration (mg/L) of carbon (C). Carbon is the common component of all organic matters and the main element of organic matters. The higher TOC value of water means the higher content of organic matters in water. Therefore, TOC can be used as an index to evaluate organic pollution of water quality. It excludes other elements, such as N, S or P elements with high content contained in organic compounds, which also participate in the oxidation reactions during the combustion oxidation process. The TOC calculated by C cannot reflect the content of this part of organic matters.

TOC is measured by instrument method. According to different working principles, it can be divided into combustion oxidation-non dispersive infrared absorption method, conductivity method, and wet oxidation-non dispersive infrared absorption method, etc. The combustion oxidation-non dispersive infrared absorption method has a simple process, good reproducibility and high sensitivity, and is widely used at home and abroad. The determination of TOC by combustion oxidation-non dispersive infrared absorption method can be divided into subtraction method and direct method. Because some carbon containing organic compounds are not easily burned and oxidized at high temperatures, the measured TOC value is often slightly lower than the theoretical value.

TOC tester has the advantages of simple process, good reproducibility, high sensitivity, high stability and reliability, no consumption of chemicals, no secondary pollution, measuring all organic carbon content and so on. Therefore, TOC is the best comprehensive index to monitor organic pollutant emission.

Therefore, it is of great significance to master the nondispersive infrared absorption method for the determination of TOC in water.

1 Experimental purpose

(1) Master the basic principle and operation procedure of nondispersive infrared absorption method to determine TOC in water.

(2) Understand the impact of TOC content on water quality assessment.

2 Experimental principle

2.1 Determination of TOC by subtraction method

The sample, together with purified air (dried and removed with carbon dioxide), is introduced into a high-temperature combustion tube (900℃) and a low-temperature reaction tube (160℃), respectively. The water sample through the high-temperature combustion tube is catalyzed and oxidized by high temperature, so that both organic compounds and inorganic carbonates are converted into carbon dioxide. The water sample through the low temperature reaction tube is acidified to decompose the inorganic carbonates into carbon dioxide, and the carbon dioxide generated is introduced into the nondispersive infrared detector in turn. A certain wavelength of infrared is selected by carbon dioxide to absorb. And in a certain concentration range, the intensity of carbon dioxide absorption of infrared rays is proportional to the concentration of carbon dioxide. Therefore, the total carbon (TC) and inorganic carbon (IC) of water samples can be quantitatively determined.

2.2 Determination of TOC by direct method

The water sample is acidified and aerated, and the inorganic carbonate is decomposed into carbon dioxide and removed. Then, it is introduced into the high-temperature combustion tube, which can directly determine its TOC.

3 Experimental instruments and reagents

3.1 Instruments

(1) Nondispersive infrared absorption TOC analyzer. Working conditions: ambient temperature: 5-35℃; operating voltage: rated voltage of the instrument, AC; total carbon combustion tube temperature selection: 900℃; inorganic carbon reaction tube temperature control: 160℃±5℃; carrier gas flow: 180mL/min; microsyringe: 50.00μL, colorimetric tube with plug: 10mL.

(2) General laboratory instruments.

3.2 Reagents

Unless otherwise indicated, all reagents are analytical pure, and the water used is distilled water without CO_2.

(1) Distilled water without carbon dioxide: The redistilled water will be boiled in a beaker to evaporate (10% evaporation). After slight cooling, it is put into a tubualted bottle inserted with a soda lime tube for use.

(2) Potassium hydrogen phthalate ($KHC_8H_4O_4$): GR.

(3) Anhydrous sodium carbonate (Na_2CO_3): GR.

(4) Sodium bicarbonate ($NaHCO_3$): GR, stored in a desiccator.

(5) Organic carbon standard stock solution (400mg/L). 0.8500g potassium hydrogen phthalate (dried for 2h at 110-120℃ in advance, cooled to room temperature in a desiccator) is weighed, dissolved in water, transferred to a 1000mL volumetric bottle, diluted

with water to the line, and mixed well. It can be stored at low temperature (4℃) for 48 days.

(6) Organic carbon standard solution (80mg/L). 10.00mL organic carbon standard stock solution is accurately transferred. Then it is placed in a 50mL volumetric bottle, diluted with water to the standard line, and mixed well. The solution should be prepared when used.

(7) Inorganic carbon standard stock solution (400mg/L). 1.400g sodium bicarbonate (pre-dried in a desiccator) and 1.770g anhydrous sodium carbonate (pre-dried at 105℃ for 2h, placed in a desiccator, cooled to room temperature) are weighed, dissolved in water, transferred to a 1000mL volumetric bottle, diluted with water to the line, and mixed well.

(8) Inorganic carbon standard solution (80mg/L). 10.00mL inorganic carbon standard stock solution (7) is accurately added into a 50mL volumetric bottle, diluted with water to the line, and mixed well. The solution should be prepared when used.

4 Experimental procedure

4.1 Sample collection and storage

After the water sample is collected, it must be stored in a brown glass bottle. The water sample can be stored for 24h at normal temperature. If it cannot be analyzed in time, the pH of the water sample can be adjusted to 2 or less with sulfuric acid, and the sample should be refrigerated at 4℃, which can be stored for 7 days.

4.2 Instrument debugging

Debug TOC analyzer and recorder according to instructions. Select the sensitivity, measurement range, total carbon combustion tube temperature and carrier gas flow. The instrument is energized and preheated for 2h until the baseline on the recorder is stable when infrared analyzer output.

4.3 Elimination of interference

When the content of common co-existing ions in water samples exceeds the allowable value of interference, the absorption of infrared rays will be affected. In this case, the water sample must be diluted with distilled water without CO_2. The analysis is performed again when the content of all co-existing ions is lower than the allowable concentration of interference.

4.4 Subtraction method

Acidified water samples should be neutralized to neutral with sodium hydroxide solution before determination. $20.9\mu L$ mixed water sample is accurately injected into total carbon combustion tube and inorganic carbon reaction tube successively with $50.00\mu L$ microsyringes. The corresponding absorption peak height that appears in the recorder could be measured.

4.5 Direct method

About 25mL acidified water sample is added into a 50mL beaker. It is stirred vigorously on a magnetic stirrer for a few minutes or N_2 without CO_2 is injected into the beaker to remove the inorganic carbon. $20\mu L$ water sample without inorganic carbon is injected into the total carbon combustion tube. The absorption peak height appears in the recorder could be measured.

4.6 Blank experiment

According to the above steps, a blank experiment is performed. And the sample is replaced by $20.0\mu L$ water.

4.7 Calibration curve drawing

0.00, 0.50, 1.50, 3.00, 4.50, 6.00 and 7.50mL organic carbon standard solution and inorganic carbon standard solution are added respectively into a set of seven 10mL colorimetric tubes with plugs. They are diluted with distilled water to line and mixed well. 0.0, 4.0, 12.0, 24.0, 36.0, 48.0 and 60.0mg/L organic and inorganic carbon standard series solutions are prepared. Then, according to the above steps, the measured absorption peak height of the standard series solution is subtracted from the absorption peak height of the blank test to obtain the corrected absorption peak height. The calibration curves of organic carbon and inorganic carbon are drawn from the concentration of the standard series solution and the corresponding correction absorption peak height. The linear regression equations of calibration curves can also be calculated according to the method of linear regression equation.

5 Data processing

5.1 Subtraction method

According to the correction value of absorption peak height of the tested samples minus absorption peak height of the blank, the mass concentration of total carbon (TC, mg/L) and inorganic carbon (IC, mg/L) can be calculated from the calibration curve regression equations or obtained from the calibration curves. The difference value of mass concentration between total carbon and inorganic carbon is the mass concentration of total organic carbon (TOC, mg/L) of the sample.

$$TOC(mg/L) = TC(mg/L) - IC(mg/L)$$

5.2 Direct method

According to the correction value of absorption peak height of the tested samples minus absorption peak height of the blank, the mass concentration of total carbon (TC, mg/L) can be calculated from the calibration curve regression equation or obtained from the calibration curve, which is the mass concentration of total organic carbon (TOC, mg/L) of the sample.

$$TOC(mg/L) = TC(mg/L)$$

The volume of the sample is $20.0\mu L$ and the result is expressed as one decimal place.

6 Questions

(1) Try to analyze and discuss the main sources of total organic carbon in water.

(2) What factors can interfere with the determination of TOC in water? And how to eliminate them?

Section 3 Soil environmental chemistry experiments

Experiment 16 Determination of lead in soil by graphite furnace atomic absorption spectrometry

1 Experimental purpose

(1) Understand the principles and characteristics of graphite furnace atomic absorption spectrometry.

(2) Master the operation of graphite furnace atomic absorption spectrometer.

(3) Be familiar with pretreatment of soil samples.

2 Experimental principle

Lead is a cumulative harmful element that can cause damage to the nervous system, digestive system and hematopoietic system when ingested in excess. The mineral lattice of soil is completely destroyed by the degradation method of hydrochloric acid-nitric acid-hydrofluoric acid-perchloric acid. All elements to be measured in samples will enter the test solution. The test solution is injected into the graphite furnace, and the coexisting matrix components are vaporized and removed through pre-set drying, ashing, atomization and other procedures. At the same time, the lead compounds are dissociated into the ground state atomic vapor at the high temperature of the atomization stage, and the characteristic spectral lines emitted by the hollow cathode lamp are selectively absorbed. The absorbance of lead in the test solution is determined by background deduction under the optimum determination condition.

3 Experimental instruments and reagents

3.1 Instruments

Graphite furnace atomic absorption spectrophotometer (with background deduction device), lead hollow cathode lamp, argon cylinder.

3.2 Reagents

(1) Concentrated hydrochloric acid, concentrated nitric acid, hydrofluoric acid, perchloric acid. All of them are guarantee reagents.

(2) Nitric acid solution, prepare as 1+5.

(3) Diammonium hydrogen phosphate (guarantee reagent) aqueous solution, the mass fraction is 5%.

(4) Soil samples.

(5) Lead standard reserve solution, 0.500mg/mL. Accurately weigh 0.500g spectral pure metal lead in a 50mL beaker, add 20mL nitric acid solution, and dissolve it slightly by heating. After cooling, transfer it to a 1000mL volumetric flask, dilute it with water to the

marked line, and shake it well.

(6) Lead standard liquid, 250μg/L. It is prepared by diluting lead standard reserve solution with nitric acid solution step by step before use.

4 Experimental procedure

(1) Accurately weigh and collect soil samples after pretreatment 0.1-0.3g (accurate to 0.0002g) in a 50mL teflon crucible, and add 5mL concentrated hydrochloric acid after wetting soil samples with water. Samples are initially decomposed by low temperature heating on the electric heating plate of the fume hood. When the solution evaporate to about 2-3mL, remove and let it cool down, add 5mL concentrated nitric acid, 2mL hydrofluoric acid, 2mL perchloric acid, and heat it at medium temperature for about 1h on the electric heating plate after covering. Take off the lid, continue heating to remove silicon. To achieve a good effect of silicon removal, should often shake crucible. When the sample is heated to emit thick perchlorate white smoke, cover it, the black organic carbide can be fully decomposed. After the black organics on the crucible disappear, the lid is removed to drive away the white smoke and steam sample until the contents become sticky. Depending on the situation of digestion, 2mL concentrated hydrochloric acid, 2mL hydrofluoric acid and 1mL perchloric acid can be added to repeat the digestion process. When the white smoke is basically exhausted again and the contents are viscous, take off the lid and let it slightly cool, rinse the crucible cover and the inner wall with water, and add 1mL nitric acid solution to dissolve the residue. The liquid is then transferred to a 25mL volumetric flask, and 3mL diammonium phosphate solution is added, then the volume is fixed after cooling, and the mixture is shaken well for measurement.

Due to the variety of soil and the difference of organic matter content, the amount of various acids can be increased or decreased depending on the situation of digestion. Soil digestion solution should be white or light yellow with no sediment.

(2) Adjust the instrument to the best working condition according to the manual and measure the laser degree of the test solution. Instrument measurement conditions refer to Table 4-3-1.

Table 4-3-1 Instrument measurement conditions

Determine wavelength/nm	283.3	Atomization/(℃/s)	2000/5
Passband width/nm	1.3	Clear/(℃/s)	2700/3
Lamp current/mA	7.5	Argon flow rate/(mL/min)	200
Dry/(℃/s)	80-100/20	Whether stop the gas at the atomization stage	Is
Ashing/(℃/s)	700/20	Sample volume/μL	10

(3) Blank experiment. Replace the sample with water and proceed with steps (1) and (2).

(4) Calibration curve. Use pipette to accurately transfer lead standard liquid 0.00mL, 0.50mL, 1.00mL, 2.00mL, 3.00mL, 5.00mL into a 25mL volumetric flask. Add 3.0mL diammonium hydrogen phosphate solution and dilute with nitric acid solution to volume. The standard solution contains 0μg/L, 5.0μg/L, 10.0μg/L, 20.0μg/L, 30.0μg/L, and 50.0μg/L lead. Determine the absorbance of the standard solution from low to high concentration as described in step (2). Plot the lead standard curve with the absorbance that

subtracts the blank absorbance and the corresponding element content ($\mu g/L$).

(5) According to the absorbance of the sample after deducting the blank absorbance, the lead concentration in the sample to be tested can be found on the standard curve.

5 Data processing

The mass fraction of lead in soil samples W ($\mu g/kg$) is calculated according to equation (4-3-1).

$$W = \frac{cV}{m(1-f)} \quad (4\text{-}3\text{-}1)$$

Where, c——Content of lead found on the standard curve according to the absorbance of the test solution minus the absorbance of the blank, $\mu g/L$;

V——volume of sample solution, mL;

m——mass of the sample, g;

f——moisture content of the sample, %.

Determination of soil moisture content f. 5-10g (accurate to 0.01g) air-dried soil samples passed through a 100-mesh sieve are put into a lead box or weighing bottle and baked in a 105℃ oven for 4-5h to constant weight. Moisture content f (%) of air-dried soil samples expressed in percentage is calculated according to equation (4-3-2).

$$f = \frac{W_1 - W_2}{W_1} \quad (4\text{-}3\text{-}2)$$

Where, f——moisture content of soil sample, %;

W_1——mass of soil sample before drying, g;

W_2——mass of soil sample after drying, g.

6 Questions

(1) What are the main parts of the graphite furnace atomic absorption spectrometer? What are the functions of each part?

(2) What is the significance of measuring the absorbance of blank solution in the experiment?

(3) What are the advantages and disadvantages of graphite furnace atomic absorption spectrometry compared with flame atomic absorption spectrometry?

7 Notes

(1) Hydrofluoric acid and perchloric acid used in the digestion process are explosive. The whole digestion process must be carried out in a fume copboard.

(2) The temperature of the electric heating plate should not be too high, otherwise it will make the teflon crucible deformation.

Experiment 17 Chemical remediation of soil contaminated by heavy metals (EDTA leaching of copper from soil)

Soil pollution is extremely harmful to human beings. It not only directly leads to the degradation of land quality and affects the safety of agricultural products, but also affects human health through the food chain. It also poses hazards to human living environment on multiple levels through the pollution of groundwater and the transfer of pollution. In view of the seriousness of soil pollution, the remediation of contaminated soil has received great attention in the world and become a hot spot of environmental research at home and abroad.

1 Experimental purpose

(1) Master the general methods of soil pretreatment.

(2) Learn the principle of heavy metal polluted soil remediation by leaching.

(3) Explore the optimal condition for EDTA to remediate heavy metal polluted soil and evaluate the remediation effect.

2 Experimental principle

Soil leaching method is a kind of soil remediation method. High-energy contact happens between the soil and eluent, and organic and inorganic pollutants are removed from sludge and sediment. It includes more than physical and chemical mechanisms of the repair process. It can also realize the separation, isolation, or harmless transformation of hazardous substances. Dealing with eluent containing pollutants can reduce ectopic repair cost of collection, transport and recovery, and has great potential in the repair of organic and inorganic pollution, and economically also has certain advantages over other methods, thus attracts much attention in the world. To repair heavy metal pollution by leaching, inorganic acids, organic acids with low relative molecular weight, or appropriate complexing agents such as EDTA and NTA are usually added in the leaching solution to increase the mobility of metals, so as to achieve the purpose of cleaning the soil.

The structure of sodium ethylenediamine tetraacetic acid is as follows:

$$\begin{matrix} NaOOCH_2C \\ NaOOCH_2C \end{matrix} \!\!\! > \!\! N-H_2C-CH_2-N \!\! < \!\!\! \begin{matrix} CH_2COOH \\ CH_2COONa \end{matrix}$$

It has four oxygen atoms giving away electron pairs and two nitrogen atoms giving away electron pairs, so it can form stable chelates with most metal ions. The action equation of Cu^{2+} and ethylenediamine tetraacetic acid (EDTA) is:

$$Cu^{2+} + Y^{4-} = CuY^{2-}$$

Cu^{2+} and Y^{4-} chelate to form five atomic rings, which is the main reason why chelates are especially stable.

The leaching of Cu by EDTA is equivalent to the competitive equilibrium effect of EDTA on Cu in soil.

3 Experimental instruments and reagents

3.1 Instruments

Plastic centrifuge tube, pipette and pipettor, shaker, volumetric flask, centrifuge, acidity meter, atomic absorption meter.

3.2 Reagents

(1) $CuCl_2$ standard solution: Accurately weigh 26.85mg $CuCl_2 \cdot H_2O$, add 1-2mL dilute hydrochloric acid (0.1mol/L) to dissolve to prevent the hydrolysis of Cu^{2+}, transfer the solution to a 1L volumetric flask, set the volume to the scale mark with deionized water, and shake well.

(2) EDTA aqueous solution: Weigh 18.61g EDTA disodium salt, dissolve it with 500mL deionized water, adjust the pH to 5.0 with 1mol/L NaOH, transfer the solution to a 1L volumetric flask, set the volume to the scale mark with deionized water, and shake well.

(3) 0.05mol/L $NaNO_3$ solution: Accurately weigh 4.25g $NaNO_3$, transfer it to a beaker, add 500mL deionized water, stir it until completely dissolved, transfer the solution to a 1L volumetric flask, set the volume to the scale mark with deionized water, and shake well.

(4) HCl: GR.

(5) HNO_3: GR.

(6) $HClO_4$: GR.

(7) NaOH: GR.

4 Experimental procedure

4.1 Preparation of contaminated soil

The soil samples collected from the field are poured onto plastic film or paper. The soil blocks are crushed in the semi-dry state to remove residual roots, stones and other debris. Spread into a thin layer in the shade and slowly dried. After air drying, the soil samples are crushed with a hard rod and passed through a 2mm nylon sieve. The percentage of sand and gravel larger than 2mm in the whole soil sample should be calculated. The soil smaller than 2mm is repeatedly sampled with quartering method. The samples are further refined with agate mortar and screened with 100mesh sieve. The whole bottle of soil samples are ready for use.

4.2 Determination of Cu content in contaminated soil

4.2.1 Digestion of soil samples

Accurately weigh 1.000g soil samples (3) and soil standard samples (3) in 100mL beakers, moisten them with a small amount of deionized water, slowly add 5mL aqua regia, and cover with the surface dish. At the same time, make a reagent blank, heat the beaker on the electric stove in the fume hood, start with low temperature then slowly raise the temperature, and keep it in a slightly boiling state to make it fully decompose. Pay attention to the digestion temperature should not be too high to prevent the sample from spattering. Wear gloves and goggles when operating. When the reaction is completed and most of the organic matters have decomposed, remove the beakers for cooling, add about 3mL perchloric acid along the wall of beakers, and continue to heat and decompose until white smoke appears and samples become gray. Remove surface dish and excessive perchlorate, steam the samples to nearly dry, and take off and cool them. Add 5mL 1% dilute nitric acid in it and heat, cool and filter it with medium speed quantitative filter paper to a 50mL volumetric flask. Filter residue is washed with nitric acid of 1%, diluted to volume, and finally shaked well for test.

4.2.2 Preparation of standard curve solution

Take six 50mL volumetric bottles, add five drops of 1:1 hydrochloric acid, successively add 0.0mL, 2.00mL, 4.00mL, 6.00mL, 8.00mL, 10.00mL copper standard solution with concentration of 10mg/L, dilute them with deionized water to scale, and shake well.

4.2.3 Determination

The standard solution and digestive solution are directly injected into the air-acetylene flame. The absorption value is measured with atomic absorption detector.

4.3 Determine the relationship between elution ratio and pH

Take fifiteen 50mL plastic centrifuge tubes and weigh 0.5000g copper contaminated soil respectively into these tubes.

Add 600.0mL deionized water to a 1000mL beaker. Add suitable amount of EDTA, adjust the mole ratio of EDTA and Cu in the system is 4:1. The content of Cu in the solution is calculated according to the amount determined by preparing standard curve. Assume that measured content of Cu in the contaminated soil is c, and 30.0mL drench lotion will be introduced in centrifuge tube. The concentration of Cu in the leaching system is as follows:

$$\text{concentration of Cu(mol/L)} = 0.500c/(63.5 \times 0.030) \tag{4-3-3}$$

Sodium nitrate is used to adjust the ionic strength and the concentration of sodium nitrate in EDTA solution is 0.01mol/L. HCl and NaOH are used to adjust the pH of the solution. The pH of the solution is first adjusted to 4, and 30mL solution is transfered into three centrifuge tubes. Then adjust the pH of the solution to 5, transfer 30mL solution into three centrifuge tubes respectively, and so on, and make four pH points (4, 5, 6, 7) respectively. Now shake the centrifuge tubes with added centrifugal agents in the shaker for 24h at 250 times/min, then take them out. Centrifuged tubes at 2400r/min for 15min, the supernatant is determined with 5mL by AAS, and the remaining solution is measured for pH.

4.4 Determine the relationship between elution ratio and molar ration

Take fifiteen 50mL plastic centrifuge tubes and weigh 0.5000g copper contaminated soil into them respectively.

Take five 250mL beakers, add 150mL deionized water, and then add an appropriate amount of EDTA, so that the molar ratio of EDTA and Cu in each beaker is 2:1, 4:1, 6:1, 8:1, 10:1 respectively. The determination method of the amount of Cu and EDTA is the same as 4.3. Adjust the ionic strength of solution with sodium nitrate, so that the concentration of sodium nitrate in each EDTA solution is 0.01mol/L.

According to the optimal pH determined in 4.3, adjust the pH of EDTA aqueous solution eluent in five beakers. After that, take three 30mL eluent in three centrifuge tubes with weighed copper contaminated soil. The centrifuge tubes are placed in a shaker, shaken and vibrated at 250 times/min for 24h. It was removed and centrifuged for 15min at 4500r/min. The supernatant is determined with 5mL by AAS, and the remaining solution is measured for pH.

4.5 Determine the relationship between elution ratio and time

Basic steps are the same as those mentioned above. According to the above conditions, the molar ratio, ionic strength and pH are adjusted. A total of thirteen points are made, three parallel samples for each point. From 0h to 24h, samples are taken every two hours and centrifuged, the supernatant is determined with 5mL by AAS, and the remaining solution is measured for pH.

5 Data processing

5.1 Content of copper in contaminated soil samples

5.1.1 Draw standard curve

The measured absorption value of the standard solution is plotted against the concentration (mol/L).

5.1.2 Calculate the content of copper in contaminated soil samples

According to the measured absorption values (if the reagent blank has absorption, the blank absorption value shall be deducted), corresponding mass concentration M (mg/mL) can be obtained by the standard curve. The mass fraction of copper in soil is calculated according to equation (4-3-4).

$$\text{mass fraction of Cu} = \frac{MV}{m} \times 10^{-3} \quad (4\text{-}3\text{-}4)$$

Where, M——the corresponding mass concentration obtained on the standard curve, mg/mL;

V——constant volume, mL;

m——soil sample mass, g.

5.2 Relationship between elution ratio and pH

5.2.1 Draw standard curve

The measured absorption value of the standard solution is plotted against the concentration (mol/L).

5.2.2 Draw the relation diagram between elution ratio and pH

The concentration of copper (30mL solution in the centrifuge tube) in the leaching system is calculated according to equation (4-3-3).

According to the measured absorption values, corresponding concentration C_f (mol/L) can be obtained by the standard curve. Then the elution ratio of Cu is calculated according to equation (4-3-5).

$$\text{Elution ratio} = \frac{C_f}{C} \quad (4\text{-}3\text{-}5)$$

Figure out the relation between elution ratio and pH by drawing the elution ratio diagram at each pH point calculated.

5.3 The relationship between elution ratio and mole ratio

5.3.1 Draw standard curve

The measured absorption value of the standard solution is plotted against the concentration (mol/L).

5.3.2 Draw the relation diagram between elution ration and molar ratio

C, C_f and elution ratio are calculated as above. The relation between elution ratio and mole ratio is found according to the mole ratio diagram of elution ratio calculated at each mole ratio point.

5.4 Relation between elution ratio and time

The measured absorption value of the standard solution is plotted against the concentration (mol/L).

C, C_f and elution ratio are calculated as above. Draw the relation diagram between elution ratio and time, and find out the relation between elution ratio and time.

6 Questions

(1) Why should pH be measured again after soil has been leached?

(2) How to recycle the eluent during in-situ soil remediation?

(3) How to reuse the eluent?

7 Notes

(1) Chemical treatment and remediation of soil contaminated by heavy metals means using chemical reagents, chemical reactions or chemical principles to reduce the migration and biological availability of heavy metals in the soil, and reduce or even remove heavy metals in the soil, so as to achieve the purpose of treatment and remediation of contaminated soil. It includes leaching method, curing method, applying improver method and electrochemical method.

① Leaching method is the leaching of contaminated soil with eluent, also known as soil washing or extraction method. Operations are divided into local leaching and removal of soil cleaning. In most cases, in order to improve elution efficiency, chemical additives need to be added to the eluent. Low cost, biodegradable and non-soil pollution chemicals are usually selected. The most commonly used are acids and chelators. The latter are more suitable for local leaching of soil because they are less harmful to the environment. EDTA is the most effective extractant for heavy metals.

② Curing method is to solidify and immobilize heavy metals in the soil by physical methods to reduce their harm. Since cement is cheap and easy to obtain, the technology of stabilization/solidification with cement as binder has been used in the final treatment of inorganic, organic and nuclear wastes abroad, but it is not applicable to the on-site remediation of contaminated farmland.

③ Applying improver method regulates soil environment, controls reaction conditions, changes the configuration, water-solubility, mobility and biological effectiveness of pollutants, makes the pollutants passivated, and reduces their harm to ecological environment. It includes precipitation, chemical reduction method, adsorption method, antagonistic action, and improvement of organic matter.

④ Electrochemical remediation is an in-situ remediation technology that is being developed to remove soil heavy metals and radioactive elements. The principle is as follows: under the action of a certain current and voltage, different ions can migrate to the opposite electrode under the action of battery and electromigration, hydrogen ions and metal ions move to the cathode directionally and dissolve metal ions in the soil at the same time. This method is restricted by the nature of soil, and is not suitable for soil with high permeability and poor conductivity and sandy soil.

(2) In the process of soil digestion, the temperature should be carefully controlled (the reactant is easy to overflow or carbonize when the temperature rises too fast). Most organic matters in the soil should be digested and cooled before perchloric acid is added.

(3) This experiment mainly discusses the mechanism of chemistry, without too much involving the dynamic process. When studying the soil in-situ remediation, we can first explore the dynamic conditions through the original soil column experiment.

(4) Subsequent treatment and recycling of the eluent are very critical parts of in-situ remediation, and dealing with this problem is very helpful to the effect and cost of remediation.

(5) Plants can absorb copper from the soil, but the accumulation of copper in crops is not significantly related to the total copper in the soil, but closely related to the content of available copper (water-soluble copper and exchangeable copper that can be directly absorbed and utilized by plants).

Experiment 18 Determination of soil organic matter

Soil organic matter refers to organic compounds containing carbon in soil. Soil organic matter comes from a wide range of sources, such as plant residues, animal residues, microbial residues, excreta and secretions, waste water and waste residues. The content of soil organic matter varies greatly from soil to soil, with high content of over 20% or 30% (such as peat soil, some fertile forest soil, etc.) and low content of less than 1% or 0.5% (such as desert soil and aeolian sandy soil, etc.). In soil science, the soil that contains organic matter more than 20% in the cultivated layer is generally called organic matter soil, and the soil that contains organic matter less than 20% is called mineral soil. In general, the soil organic matter content of cultivated layer is usually above 5%.

Soil organic matter can be determined by methods as follows:

(1) Visual colorimetry. The soil organic matter is oxidized with potassium dichromate solution by taking glucose solution as the standard substance for reference, and the color of the oxidized solution is linearly related to the content of organic matter. Results can be obtained by direct visual colorimetry.

(2) Calcination method. Measure weight loss of soil caused by burning carbon in soil organic matter. Soil samples removed from hygroscopic water at 105℃ are weighed and burned for two hours at 350-1000℃ before weighing. The difference between the two weights is the weight of organic carbon in the soil sample.

(3) Photometric method is spectrophotometric determination of soil organic matter with ferrous sulfate solution as standard solution.

(4) Potassium dichromate volumetric method—direct heating digestion method. Under heating conditions, excessive $K_2CrO_7 \cdot H_2SO_4$ solution is used to oxidize the organic matter in the soil, and $FeSO_4$ standard solution is used to titrate the remaining K_2CrO_7 to determine the content of organic matter in the soil.

Visual colorimetry is simple and easy to use, and can get the analysis results quickly, but the error of the measured results is large. The method of cauterization is simple, but its cauterization time is long, the cauterization temperature is high, the error of measured results is large, and it has some limitations. But these two methods are convenient, quick and have little pollution. Direct heating digestion method for oil bath heating has improved the heating condition. Photometric method changes the titration of potassium dichromate volumetric method to photometry. The determination results of two methods have good data correlation and accuracy. However, if you have large quantities of samples, direct heating digestion method is time-consuming, and photometric method can ensure the accuracy of the determination results while achieving rapid determination in large quantities.

Soil organic matter is an important source of various nutrients in soil, especially nitrogen. In general, the content of soil organic matter is an important index of soil fertility.

1 Experimental purpose

Master potassium dichromate volumetric method to determine the soil organic matter.

2 Experimental principle

The potassium dichromate volumetric method is commonly used in soil organic matter determination. Although the oxidants, concentrations or specific conditions are slightly different in various volumetric methods, the basic principle is oxidizing soil organic carbon with excessive potassium dichromate (K_2CrO_7) solution in the presence of sulfuric acid, and then titrating the remaining potassium dichromate with ferrous sulfate solution. The content of organic carbon in soil is calculated indirectly according to the amount of potassium dichromate consumed, and then the content of organic matter in soil is calculated according to the proportion of organic matter and organic carbon in soil (i. e. conversion factor).

K_2CrO_7 does not completely oxidize the organic compounds in the soil, so a correction factor is required to correct the content of unreacted organic carbon. It is generally believed that the organic carbon oxidized by this method is only 90% of the actual content, i. e., the correction factor is 1.1. The specific reaction process of this method is as follows:

Oxidation reaction:
$$2K_2Cr_2O_7 + 8H_2SO_4 + 3C \longrightarrow 2K_2SO_4 + 2Cr_2(SO_4)_3 + 3CO_2 + 8H_2O$$

Titration reaction:
$$K_2Cr_2O_7 + 6FeSO_4 \longrightarrow K_2SO_4 + Cr_2(SO_4)_3 + 3Fe_2(SO_4)_3 + 7H_2O$$

During titration, the phenanthroline oxidation-reduction indicator is used to indicate the end point of titration. The color change of the solution during the whole reaction is as follows. The titration start with orange of potassium dichromate, and the green of Cr^{3+} appears in the process of titration. Due to the formation of trivalent iron at the same time, the overall color of the solution is blue-green. When excessive potassium dichromate as strong oxidant is consumed, standard ferrous sulfate solution is half a drop too much, and the ferrous ion will combine with the phenanthroline indicator to appear brownish red, indicating that the titration end has been reached.

3 Experimental instruments and reagents

3.1 Instruments

Electronic balance, glass tube, triangular bottle, small funnel, 5mL pipette for two, oven, 360℃ thermometer for one, glass rod for one, 25mL piston burette for one.

3.2 Reagents

(1) Silica: in powder form.

(2) O-phenanthroline indicator: weigh 1.485g o-phenanthroline in 100mL distilled water containing 0.695g ferrous sulfate. This indicator is perishable and should be sealed and stored in a brown bottle for later use.

(3) 0.068mol/L $K_2Cr_2O_7$ solution: weigh 20g analytical pure potassium dichromate and dissolve it in 500mL distilled water. After cooling, adjust the volume to 500mL and put it into a brown reagent bottle.

(4) 0.1mol/L $FeSO_4$ solution: weigh 28g chemical pure ferrous sulfate, dissolve it in 600-800mL water, add 20mL concentrated sulfuric acid, stir evenly, add water to 1L, store it in a brown bottle, and titrate with 0.068mol/L potassium dichromate every day when using.

(5) 0.068mol/L potassium dichromate standard solution: take a certain amount of samples, dry them at 130℃ for 1.5h, weigh 2g in a small amount of water, and finally dilute it with water to 100mL in a volumetric flask. Store in a bottle.

4 Experimental procedure

4.1 The samples

Select representative air-dried soil samples, remove organic residues such as plant roots and leaves with tweezers. Press the soil clods until they can pass through a 1mm sieve, and thoroughly mix and grind them until they can pass through a 0.25mm sieve. Put them into a grinding bottle for later use. The soil sample of about 0.1-0.5g is weighed in the weighing tube with the decrement method, which is denoted as m. Then transfer the soil sample from the weighing tube to the dried triangular bottle.

4.2 Add oxidant

Draw 5mL 0.068mol/L potassium dichromic solution and 5mL concentrated sulfuric acid with pipettes into a triangular bottle with the sample, and insert a small glass funnel at the mouth of each triangular bottle.

4.3 Oven heating

The temperature of the incubator is raised to 185℃. After preheating, the sample to be tested is heated and the solution is allowed to boil for five minutes at 170-185℃.

Note: A small number of small bubbles are produced in the triangular bottle at first, but instead of boiling, carbon dioxide is released by the oxidation of organic matter. Carbonate-rich soil, in particular, produces large amounts of carbon dioxide when heated, and it cannot be timed until it is really boiling.

4.4 The titration

Take out the triangle bottle from the oven, rinse the small funnel and the inner wall of the triangle bottle with distilled water when it is cooled. The total volume of the tested solution should be controlled at 30-35 mL. Add three drops of o-phenanthroline indicator. Titrated it with 0.1 mol/L $FeSO_4$ and recorded V_2.

4.5 Blank experiment

Without adding soil samples, other steps are the same as soil determination, but the total volume of solution before titration should be controlled at 20-25 mL. Through blank experiment, the amount V_1 of 0.1 mol/L ferrous sulfate solution required for titration of 5 mL 0.068 mol/L potassium dichromate and 5 mL concentrated sulfuric acid solution can be obtained.

5 Data processing

$$\text{mass fraction of organic carbon}(\%) = \frac{(V_1 - V_2) c \times \frac{1}{6} \times \frac{3}{2} \times 12 \times 1.1}{m \times 10^3} \times 100$$

Where, V_1——$FeSO_4$ volume for blank titration, mL;

V_2——$FeSO_4$ volume for sample titration, mL;

c——calibrated concentration of $FeSO_4$ solution, mol/L;

m——mass of air-dried soil sample, g;

12——molar mass of carbon, g/mol;

1.1——correction factor.

6 Questions

How does the calcination method determine soil organic matter? What is the scope of application?

Experiment 19 Determination of soil cation exchange capacity

Soil is an important place for the migration and transformation of pollutants in the environment. Soil colloid has adsorptive property due to its huge specific surface area and electrification. In the diffusion layer of the double electric layer of the soil colloid, the compensating ions can be exchanged with the ions with the same charge in the solution on the basis of ionic valence, which is called ion exchange. Soil's adsorbability and ion-exchange properties make it the main destination of heavy metal pollutants. Soil cation exchange performance refers to the exchange between cation in soil solution and cation in soil solid phase. It is determined by the surface properties of soil colloid. Soil colloid refers to the complex organic mineral complex formed by clay minerals and humic acid in soil, and the cations absorbed by it include K^+, Na^+, Mg^{2+}, NH_4^+, H^+, Al^{3+}, etc. Soil exchange performance is of great significance for studying the environmental behaviors of pollutants. It can regulate the concentration of soil solution, ensure the diversity of soil solution components, thus maintain the physiological balance of soil solution, and also keep various nutrients from being lost by rain. The analysis of soil exchange performance includes determination of cation exchange capacity, exchangeable cation analysis and calculation of base saturation. Cation exchange capacty (CEC) is the amount of cations that soil colloid can absorb, expressed as cmol per kilogram of soil (cmol/kg). The cation exchange capacity can be used as an index to evaluate soil fertility. Cation exchange capacity is the main source of soil buffering property, which is an important basis for improving soil and applying fertilizer rationally. Therefore, it is very important to determine the cation exchange capacity that reflects the total negative charge of soil and is the important index that characterizes soil properties. The determination of soil cation exchange capacity is affected by many factors, such as the nature of the exchangers, the concentration and pH of salt solution, and the leaching method. So the operation technique must be strictly mastered to obtain reliable results. The FAO prescripts the use of the classical neutral ammonium acetate or sodium acetate method in soil analysis for soil classification. Neutral ammonium acetate method is also a conventional analysis method used by soil and agrochemical laboratories in China, which is suitable for acidic and neutral soil. Recent soil chemistry studies have shown that, for acid and slightly acid soil in tropical and subtropical regions, the results of conventional methods differ greatly from the actual situation due to the high pH value and ionic strength of the leaching solution, and the results are much higher than the actual situation. The new method saturates the soil with $BaCl_2$, balances the soil with $BaCl_2$ solution at a concentration equal to the ionic strength of the soil solution, and then uses $MgSO_4$ to exchange Ba to determine the cation exchange capacity of acid soil. Methods for determination of cation exchange capacity in calcareous soil include NH_4Cl-NH_4OAc method, $Ca(OAc)_2$ method and NaOAc method. At present, NH_4Cl-NH_4OAc method is widely used and is considered to be a better method. The results are accurate, stable and reproducible. NaOAc method is widely used in the determination of cation exchange capacity in calcareous soil and saline soil. With the development of soil analytical chemistry, the method has been developed to determine the effective cation exchange capacity in soil. Determination of cation exchange capacity by summation method is required by USDA. For highly weathered soils with variable-charge in tropical and subtropical regions, IITA recommends the determination of effective cation exchange capacity (ECEC) with summation method. Recently, international standard methods for determination of effective cation exchange capacity (ECEC or Q^+, E) and potential cation exchange capacity (PCEC or Q^+, P) in soil have been proposed, such as ISO 11260: 2018 and ISO 13536: 1995 (P), which are suitable for various soil types.

1 Experimental purpose

(1) Deeply understand the connotation of soil cation exchange capacity and its environmental chemical significance.

(2) Master the determination principle and method of soil cation exchange capacity.

2 Experimental principle

In this experiment, the rapid method is used to determine the cation exchange capacity. The various cations present in the soil can be equally exchanged by the cations (Ba^{2+}) in some neutral salt ($BaCl_2$) solutions (Fig. 4-3-1). Because there is an exchange equilibrium in the reaction, the exchange reaction is not actually complete. When the concentration of the exchangers in the solution is increased and the number of exchange times is increased, the exchange reaction can be nearly complete. The nature of the exchange ions and the physical state of the soil also affect the degree of exchange reaction. Then Ba^{2+} in the soil is exchanged with strong electrolyte (sulfuric acid solution). Because of the formation of barium sulfate precipitation and the strong exchange adsorption capacity of hydrogen ions, the exchange reaction is basically complete. In this way, by measuring the change of sulfuric acid content before and after the exchange reaction, the amount of sulfuric acid consumed can be calculated, and then the cation exchange capacity can be calculated. Measured values of ion exchange capacity in different ways vary greatly, and the method should be indicated when reporting and applying the results.

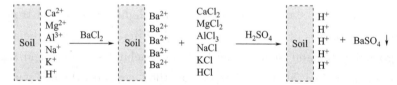

Fit. 4-3-1 Cation exchange diagram

3 Experimental instruments and reagents

3.1 Instruments

Centrifuge, centrifuge tube of 100mL, conical flask of 100mL, measuring cylinder of 50mL, pipettes of 10mL and 25mL, basic burette of 25mL.

3.2 Reagents

(1) Barium chloride solution: 60g barium chloride ($BaCl_2 \cdot 2H_2O$) is dissolved in water and transferred into a 500mL volumetric flask for constant volume with deionized water.

(2) 0.1% phenolphthalein indicator: 0.1g phenolphthalein is dissolved in 100mL alcohol.

(3) Sulfuric acid solution (0.1mol/L): Remove 5.36mL concentrated sulfuric acid to a

1000mL volumetric flask and dilute it to scale with water.

(4) Standard sodium hydroxide solution (0.1mol/L): Weigh and dissolve 0.2g sodium hydroxide in 500mL boiled but cooled distilled water. The concentration needs to be calibrated.

Calibration method: Weigh two 0.5000g potassium hydrogen phthalate (dried in the oven at 105℃ in advance) in a 250mL conical flask, and add 100mL boiled but cooled distilled water to dissolve it. Then add four drops of phenolphthalein indicator, and titrate it with the prepared standard solution of sodium hydroxide to a reddish color. A blank test is performed with boiled but cooled distilled water. The blank value is deducted from the volume of NaOH solution titrating potassium hydrogen phthalate. Then the concentration of standard sodium hydroxide solution is calculated according to equation (4-3-6).

$$N_{\text{NaOH}} = \frac{W \times 1000}{(V_1 - V_0) \times 204.23} \tag{4-3-6}$$

Where, W——mass of potassium hydrogen phthalate, g;

V_1——volume of sodium hydroxide solution consumed by titrating potassium hydrogen phthalate, mL;

V_0——volume of sodium hydroxide solution consumed by the blank titration with distilled water, mL;

204.23——molar mass of potassium hydrogen phthalate, g/mol.

4 Experimental procedure

Take four 100mL centrifuge tubes and weigh them separately (accurate to 0.0001g, the same below). Two of them are added into the surface air-dried soil samples of the sewage irrigation area, and the remaining two are added into the deep air-dried soil samples and marked. Add 20mL barium chloride solution into each tube, stir with a glass rod for 4min, and centrifuge at 3000r/min until the subsoil samples are compacted. Discard the supernatant, add 20mL barium chloride solution, and repeat above operations.

Add 20mL distilled water into each centrifuge tube, stir with glass rod for 1min, centrifuge, and discard the supernatant. Weigh the centrifuge tube together with the soil sample. 25.00mL 0.1mol/L sulfuric acid solution is transferred to each centrifuge tube, stirred for 10min, placed for 20min, centrifuged and settled, and the supernatant is poured into four tubes respectively. Then transfer 10.00mL supernatant from each tube to four 100mL conical flasks. At the same time, 10.00mL 0.1mol/L sulfuric acid solution is transferred to the other two conical flasks. 10mL distilled water and one drop of phenolphthalein indicator are added into the six conical flasks respectively. Titrate the solution with standard sodium hydroxide until it turns red and did not fade for several minutes as the end point.

5 Data processing

The soil cation exchange capacity (CEC) is calculated according to equation (4-3-7).

$$\text{CEC} = \frac{[A \times 25 - B \times (25 + G - W - W_0)] \times N}{W \times 10} \times 100 \tag{4-3-7}$$

Where, CEC——soil cation exchange capacity, cmol/kg;
A—— volume of standard sodium hydroxide solution consumed by titrating 0.1mol/L sulfuric acid solution, mL;
B——the consumption volume of standard sodium hydroxide solution consumed by titrating, supernatant after centrifugal sedimentation, mL;
G——weight of the centrifugal tube together with the soil sample, g;
W——weight of empty centrifugal tube, g;
W_0——weight of soil sample measured, g;
N——concentration of standard sodium hydroxide solution, mol/L.

6 Questions

(1) Explain the difference of cation exchange capacity between the two soils.

(2) In addition to the methods used in the experiment, are there other methods that can be used to determine the soil cation exchange capacity? What are their advantages and disadvantages?

(3) The effects of soil ion exchange and adsorption on the migration and transformation of pollutants are discussed.

7 Notes

(1) The glassware used in the experiment should be clean and dry to avoid experimental errors.

(2) During centrifugation, it should be noted that the weight of the centrifuge tube in the corresponding position should be close to that of the tube, so as to avoid the occurrence of unbalanced weight.

Experiment 20 Determination of copper, zinc, lead, and cadmium in industrial solid waste by flame atomic absorption spectrometry

Solid waste refers to the solid and semi solid waste materials generated by humans in production, consumption, daily life, and other activities, commonly known as garbage. Mainly including solid particles, garbage, slag, sludge, discarded products, damaged utensils, defective products, animal carcasses, spoiled food, human and animal feces, etc.

Industrial solid waste refers to solid waste such as mining waste, beneficiation tailings, fuel waste, chemical production and smelting waste generated in industrial, transportation and other production activities, also known as industrial residue or industrial refuse, including industrial waste, debris, sludge, tailings and other waste.

The main harm of copper to the human body is that it may lead to copper poisoning, thereby affecting the metabolic function of the body and causing damage to liver and kidney function. Copper is an essential trace mineral for the human body, and when the body's copper content is within the normal range, it is usually not harmful to the human body. When the body is exposed to a large amount of copper or has abnormal copper metabolic function, it may lead to the accumulation of a large amount of copper element in the body. Copper is a kind of heavy metal, and the accumulation of a large amount of heavy metal can cause damage to liver and kidney function. Severe cases can lead to complications such as hepatitis, hypotension, coma, hemolysis, acute kidney failure, convulsions, and even death.

Excessive zinc intake can cause poisoning, such as acute zinc poisoning, with gastrointestinal symptoms such as vomiting and diarrhea. Inhalation of zinc mist in factories can cause symptoms of low fever and cold like symptoms. Chronic zinc poisoning may have symptoms such as anemia. Animal experiments can cause damage to liver and kidney function, and immunity. Some children's toys have zinc in their paint, and children like to put the toys into their mouths. Eating too much zinc can cause poisoning.

Lead is a widely present industrial pollutant that can affect the functions of the human nervous system, cardiovascular system, skeletal system, reproductive system, and immune system, causing diseases of the gastrointestinal tract, liver, kidney, and brain.

Long term consumption of cadmium contaminated food may lead to "pain sickness", that is, excessive accumulation of cadmium in the body damages the function of renal tubules, causing protein loss from the urine, leading to the formation of cartilage disease and spontaneous fractures over time.

Therefore, mastering the determination of copper, zinc, lead, and cadmium content in industrial solid waste by flame atomic absorption spectrometry is of great significance.

1 Experimental purpose

(1) Master the basic principle and operational steps of flame atomic absorption spectrometry for determining the content of copper, zinc, lead, and cadmium in industrial solid waste.

(2) Understand the impact of copper, zinc, lead, and cadmium content on environmental assessment.

2 Experimental principle

Atomic absorption spectroscopy (AAS) is established by utilizing the phenomenon that gaseous atoms can absorb a certain wavelength of light radiation, causing the electrons transition in the outer layer of the atom from the ground state to the excited state. Due to the different energy levels of electrons in various atoms, they will selectively resonate and absorb radiation of a certain wavelength, which is exactly equal to the wavelength of the excited emission spectrum of the atom. When the light with characteristic wavelength emitted by a light source passes through atomic vapor, i.e. the frequency of incident radiation is equal to the energy frequency required for electrons in the atom to transition from the ground state to a higher energy state (usually the first excited state), the outer electrons in the atom will selectively absorb the characteristic spectral lines emitted by the same element, and weaken the incident light. The degree of the characteristic spectral lines weaken due to absorption is called absorbance A, which is proportional to the content of the measured element within a linear range.

$$A = KC$$

In the equation, A is absorbance; K is a constant (all constants); C is the sample concentration.

This formula is the theoretical basis for quantitative analysis using atomic absorption spectroscopy.

Inject the test solution directly into the flame, and in the air-acetylene flame, the compounds of copper, zinc, lead, and cadmium dissociate into ground state atoms, and selectively absorb the characteristic radiation lines of the hollow cathode lamp. Measure the absorbance of copper, zinc, lead, and cadmium under given conditions.

3 Experimental instruments and reagents

3.1 Instruments

Atomic absorption spectrophotometer with copper, zinc, lead, cadmium hollow cathode lamps, acetylene cylinders or acetylene generators, and air compressors (equipped with oil, water, and dust removal devices). Instrument parameters are shown in Table 4-3-2.

Table 4-3-2　Instrument parameters

Element	Copper	Zinc	Lead	Cadmium
Detection wavelength/nm	324.7	213.8	283.3	228.8
Pass band width/nm	1.0	1.0	2.0	1.3
Flame properties	lean-burn	lean-burn	lean-burn	lean-burn
Other optional spectral lines/nm	327.4, 225.8	307.6	217.0, 261.4	326.2

3.2 Reagents

(1) Preparation of metal standard stock solution (1.000g/L): Weigh 1.000g spectral pure metal of copper, zinc, lead, and cadmium separately, dissolve them in 20mL high-grade pure nitric acid (1:1), and dilute them to a constant volume of 1000mL, respectively.

(2) Preparation of metal mixed standard solution: Prepare a mixed standard solution containing 20.00mg/L copper, 10.00mg/L zinc, 40.00mg/L lead, and 10.00mg/L cadmium by using standard stock solutions of copper, zinc, lead, and cadmium and a 2% nitric acid solution.

(3) Ascorbic acid (1%): Prepare when use.

4 Experimental procedure

4.1 Sample collection and preservation

Weigh 100g sample (on a dry basis), place it in a mixing container for leaching, and add 1L water (including the water of the sample). Fix the container used for extraction vertically on the oscillator, adjust the oscillation frequency to (110±10) times/min and amplitude to 40mm, oscillate at room temperature for 8hours, and let it stand for 16hours.

Separate the solid-liquid phase through a filtration device and immediately measure the pH value of the filtrate. The filtrate should be analyzed as soon as possible and stored for no more than one week.

4.2 Blank test

Replace the sample with water and use the same steps and reagents as the sample to determine the blank value while testing the sample.

4.3 Calibration curve drawing

Referring to Table 4-3-3, dilute the mixed standard solution with 0.2% HNO_3 solution in a 50mL volumetric flask to prepare at least 4 working standard solutions. The concentration range should include the concentration of copper, zinc, lead, and cadmium in the test sample.

Table 4-3-3 Standard series formulation and concentration

Volume of mixed standard solution added/mL	0.00	0.50	1.00	2.00	3.00	4.00
Concentration of Cd in working standard solution/(mg/L)	0.00	0.10	0.20	0.40	0.60	1.00
Concentration of Cu in working standard solution/(mg/L)	0.00	0.20	0.40	0.80	1.20	2.00
Concentration of Pb in working standard solution/(mg/L)	0.00	0.40	0.80	1.60	2.40	4.00
Concentration of Zn in working standard solution/(mg/L)	0.00	0.10	0.20	0.40	0.60	1.00

Adjust the instrument according to the selected operating parameters, zero it with nitric acid solution, and measure the absorbance of each solution in order from low concentration to high concentration. Use the measured absorbance and the corresponding concentration to draw a standard curve.

4.4 Determination

Measure the blank sample while measuring the standard solution. Determine the concentration of copper, lead, zinc, and cadmium in the test sample from the calibration curve based on the absorbance of the sample after deducting the blank absorbance. To determine the calcium residue leaching solution and reduce calcium interference, it is necessary to dilute the leaching solution appropriately. When determining the lead in chromium residue leaching solution, in addition to appropriately diluting the leaching solution, in order to prevent low lead determination results, 5mL 1% ascorbic acid is added to 50mL test solution to reduce hexavalent chromium to trivalent chromium, so as to avoid the formation of lead chromate precipitation. When the concentration of silicon in the sample is greater than 20mg/L, add 200mg/L calcium to avoid low zinc determination results.

During the process of determining the sample, it is necessary to regularly retest the blank and working standard solution to check the stability of the baseline and whether the instrument sensitivity line has changed.

5 Data processing

Calculate the concentration c of Cu, Zn, Pb and Cd in the leaching solution according to equation (4-3-8):

$$c(\text{mg/L}) = c_1 \times \frac{V_0}{V} \qquad (4\text{-}3\text{-}8)$$

Where, c_1——concentration of metal ions in the test sample, mg/L;
V_0——fixed volume during sample preparation, mL;
V——volume of test sample, mL.

6 Questions

(1) Analyze and discuss the main sources of copper, zinc, lead, and cadmium in solid waste.

(2) What factors will interfere with the determination of copper, zinc, lead, and cadmium in solid waste, and how to eliminate them?

Section 4 Ecological effects experiments of chemical substances

Experiment 21 Adsorption of phenol by sediment

Sediment/suspended particulate matter is the source and sink of pollutants in water. There are many ways for the migration and transformation of organic pollutants in water, such as volatilization, diffusion, chemical or biological degradation, etc., among which the adsorption of sediment/suspended particles has an important impact on the migration, transformation, tendency and biological effects of organic pollutants, and plays a decisive role to some extent. The adsorption of organic matter by bottom mud mainly includes distribution and surface adsorption.

Phenol is a basic raw material of chemical industry and a common organic pollutant in water. The adsorption of phenol by sediment is related to its composition and structure. The strength of adsorption can be expressed by adsorption coefficient. It is of great significance to study the adsorption of phenol by sediment to understand the environmental chemical behavior of phenol in water/sediment, and even to prevent and control water pollution.

In this experiment, phenol in water is adsorbed by sediment as adsorbent. After drawing adsorption isotherm, the adsorption constant of sediment to phenol is obtained by regression method.

1 Experimental purpose

(1) Draw the adsorption isotherm of sediment to phenol, and calculate the adsorption constant.

(2) Understand the environmental chemical significance of sediment in water and its role in water body self-purification.

2 Experimental principle

The adsorption of sediment to a series of concentrations of phenol is experimentally studied, the equilibrium concentration and the corresponding adsorption amount are calculated, and the adsorption performance and mechanism of sediment are analyzed by drawing an isothermal adsorption curve.

4-aminoantipyrine method is used to determine phenol. That is, phenol reacts with 4-aminoantipyrine to generate orange indoxol antipyrine dye with the presence of potassium ferricyanide in the medium of pH=10.0±0.2, and its aqueous solution has the maximum absorption at the wavelength of 510nm. The minimum detectable concentration of phenol is 0.1mg/L when using a 2cm covette.

3 Experimental instruments and reagents

3.1 Instruments

Constant temperature and adjustable speed oscillator, low-speed centrifuge, visible light spectrophotometer, iodine measuring bottle of 100mL, centrifuge tube of 50mL, colorimetric tube of 50mL, pipette of 0.5mL, 1.0mL, 2mL, 5mL, 10mL, 20mL.

3.2 Reagents

(1) Preparation and characterization of sediment samples: Collect the surface sediment of the river, remove sand, gravel, plant residues and other large pieces, and air dry samples at room temperature. Crush with porcelain mortar, pass 100-mesh sieve (<0.15mm), shake well, bottle, and set aside. The content of organic carbon (f_{oc}) in soil is determined by solid total organic carbon analyzer.

(2) Phenol free water: Add 0.2g activated carbon powder activated at 200℃ for 0.5h in 1L water. After full oscillation, leave it overnight. Filter it with a double-layer medium speed filter paper, or add sodium hydroxide to make the water alkaline, drop potassium permanganate solution until the water appears purple red, transfer it into a distillation bottle to heat and distill, and collect the outflow liquid for later use. Phenol free water should be used in this experiment.

Note: Phenol free water should be stored in glass bottles, and avoid contacting with rubber products (rubber plugs or latex tubes).

(3) Starch solution: Weigh 1g soluble starch, mix it into a paste with a small amount of water, add boiling water to 100mL, cool it, and store it in the refrigerator.

(4) Potassium bromate-potassium bromide standard reference solution of ($c_{1/6KBrO_3}$ = 0.1mol/L): Weigh and dissolve 2.784g potassium bromide in water, add 10g potassium bromide and dissolve it, transfer it into a 1000mL volumetric flask, and dilute it to the mark line.

(5) Standard reference solution of potassium iodate ($c_{1/6KIO_3}$ = 0.0125mol/L): 0.4458g potassium iodate, dried at 180℃ in advance, is weighed, dissolved in water, transferred into a 1000mL volumetric flask, and diluted to the mark line.

(6) Standard solution of sodium thiosulfate ($c_{Na_2S_2O_3}$ ≈ 0.0125mol/L): 3.1g sodium thiosulfate is weighed and dissolved in boiled and cooled water. 0.2g sodium carbonate is added and dilluted to 1000mL. Before use, standard reference solution of potassium iodate is used for calibration.

Calibration method: 10.0mL potassium iodate solution is placed in a 250mL iodine measuring bottle, and diluted to 100mL with water. 1g potassium iodide is added, and 5mL (1+5) sulfuric acid is added. The solution is plugged and shaken gently. Place it in a dark place for five minutes, titrate it with sodium thiosulfate solution until it appears pale yellow, add 1mL starch solution, continue titrating until the blue color just fades, and record the amount of sodium thiosulfate solution consumed. The concentration (mol/L) of sodium thiosulfate solution is calculated as equation (4-4-1).

$$c_{Na_2S_2O_3} = \frac{0.0125V_4}{V_3} \qquad (4-4-1)$$

Where, V_3——volume of sodium thiosulfate solution consumed, mL;

V_4——volume of standard reference solution of potassium iodate, mL;

0.0125——concentration of potassium iodate standard reference solution, mol/L.

(7) Phenol standard reserve solution: 1.00g colorless phenol is weighed and dissolved in water. Then transfer it to a 1000mL volumetric flask and dilute it to the mark line. The solution can be stored in refrigerator for at least one month.

Calibration method: draw 10.00mL phenol reserve solution into a 250mL iodine measuring bottle, dilute it with water to 100mL, add 10.0mL 0.1mol/L potassium bromate-potassium bromide solution, immediately add 5mL hydrochloric acid, plug the bottle, gently shake well, and place it in the dark for 10min. Add 1g potassium iodide, plug the bottle, gently shake well, and place it in the dark for 5minutes. Titrate it with 0.0125mol/L sodium thiosulfate standard solution to pale yellow appears, add 1mL starch solution, continue titrating until the blue just fades, and record the dosage. At the same time, water is used as blank test to replace phenol reserve solution, and the titration amount of sodium thiosulfate standard solution is recorded. The mass concentration of phenol reserve solution is calculated as equation (4-4-2).

$$\rho_{phenol} = \frac{(V_1 - V_2)c \times 15.68}{V} \qquad (4-4-2)$$

Where, ρ_{phenol}——mass concentration of phenol reserve solution, mg/mL;

V_1—— titration amount of sodium thiosulfate standard solution in blank test, mL;

V_2——titration amount of sodium thiosulfate standard solution when titrating phenol reserve solution, mL;

V——volume of phenol reserve liquid, mL;

c——concentration of sodium thiosulfate standard solution, mol/L;

15.68——1/6 molar mass of phenol, g/mol.

(8) Phenol standard intermediate solution (prepared on the same day of use): Appropriate amount of phenol storage solution is taken, diluted with water, and prepared into $10\mu g/mL$ phenol intermediate solution.

(9) Adsorption solution of phenol ($2000\mu g/mL$): Weigh and dissolve 2.00g colorless phenol in water, transfer it into a 1000mL volumetric flask, and dilute it to the mark line.

(10) Buffer solution (pH is about 10): Weigh and dissolve 20g ammonium chloride in 100mL ammonia water, and plug and store it in refrigerator.

(11) 2% 4-aminoantipyrine solution: Weigh and dissolve 2g 4-aminoantipyrine ($C_{11}H_{13}N_3O$) in water, dilute it to 100mL, and store it in refrigerator. It can be used for one week.

(12) 8% potassium ferricyanide solution: Weigh and dissolve 8g potassium ferricyanide $\{K_3[Fe(CN)_6]\}$ in water and dilute it to 100mL. It can be stored in refrigerator for one week.

4　Experimental procedure

4.1　Standard curve drawing

Add 0.00, 1.00, 3.00, 5.00, 7.00, 10.00, 12.00, 15.00, 18.00mL phenol standard solution with concentration of $10\mu g/mL$ into nine 50mL colorimetric tubes respectively and dilute them to the scale with water. Add 0.5mL buffer solution and mix well. At this time, the pH is 10.0 ± 0.2, and 1.0mL 4-aminoantipyrine solution is added and mixed well. Add 1.0mL potassium ferriccyanide solution, mix it thoroughly, and place it for 10min. Immediately use a 2cm covette to measure the absorbance at the wavelength of 510nm with distilled water as the reference, record the data, and draw the standard curve of absorbance to phenol content ($\mu g/mL$) after blank correction.

4.2　Adsorption experiment

Take six clean 100mL iodine measuring bottles and put about 1.0g (accurate to 0.0001g, the same below) sediment samples in each bottle. Then add $2000\mu g/mL$ adsorption solution of phenol and phenol free water according to Table 4-4-1. After sealing and shaking bottles, put them on the oscillator to oscillate at the speed of 150-175r/min for 8h at (25 ± 1.0)°C, and let them stand for 30min. Centrifuge on low speed centrifuge for 5min at the speed of 3000r/min, remove the supernatant of 10mL to 50mL volumetric flasks and set volume with distilled water to the scale, shake well, take out a few milliliters (depending on the equilibrium concentration) to 50mL colorimetric tubes, and dilute them to the scale. The absorbance is determined by the same steps as the standard curve, and the mass concentration of phenol is detected from the standard curve. Record the experimental data in table 4-4-1.

Table 4-4-1 Series of phenol addition concentration

Projects	1	2	3	4	5	6
Phenol adsorption solution volume/mL	1.0	3.0	6.0	12.5	20.0	25.0
Phenol free water volume/mL	24	22	19	12.5	5	0
Initial concentration, ρ_0/(mg·L^{-1})	80	240	480	1000	1600	2000
Volume of supernatant/mL	2.00	1.00	1.00	1.00	0.50	0.50
Dilution multiple	125	250	250	250	500	500
Absorbance						
Equilibrium concentration, ρ_e/(mg·L^{-1})						
Adsorption capacity, Q/(mg·kg^{-1})						

5 Data processing

(1) Calculation of equilibrium concentration (ρ_e) and adsorption capacity (Q), which is calculated by equation (4-4-3) and equation (4-4-4).

$$\rho_e = \rho_1 n \qquad (4\text{-}4\text{-}3)$$

$$Q = \frac{(\rho_0 - \rho_e)V}{m} \qquad (4\text{-}4\text{-}4)$$

Where, ρ_0——initial concentration, μg/mL;

ρ_e——equilibrium concentration, μg/mL;

ρ_1——measured concentration found on the standard curve, mg/L;

n——dilution multiple of solution;

V——volume of phenol solution added in adsorption experiment, mL;

m——mass of sediment sample added in the adsorption experiment, g;

Q——adsorption capacity of phenol on sediment sample, mg/kg.

(2) Draw the adsorption isothermal curve of phenol on sediment with equilibrium concentration and adsorption capacity data.

(3) By using Freundlich adsorption equation $Q = K\rho^{1/n}$, the constants K and n in the equation are obtained by regression analysis.

6 Questions

(1) What are the factors that affect the adsorption coefficient of sediment to phenol?

(2) Which adsorption equation can more accurately describe the isothermal adsorption curve of sediment to phenol?

Experiment 22 Evaporation rate of organic matter in water

Organic pollutants in water environment carry out different migration and transformation processes according to their physical and chemical properties and environmental conditions, such as volatilization, microbial degradation, photolysis and adsorption, etc. Recent studies have shown that volatilization from water into air is the main migration route of hydrophobic organic pollutants, especially high volatile organic pollutants.

The volatilization of organic pollutants in water conforms to the first-order kinetic equation, and its volatilization rate constant can be obtained through experiments. Its value is influenced by temperature, water velocity, wind speed, water composition and other factors. It is of great significance of determining the volatilization rate of organic matter in water to study its end-result in the environment.

1 Experimental purpose

Master the experimental method of determining the volatilization rate of dissolved organic matter in water.

2 Experimental principle

The volatilization of dissolved organic matter in water conforms to the first-order kinetic equation as equation (4-4-5).

$$\frac{dc}{dt} = -K_v c \quad (4\text{-}4\text{-}5)$$

Where, K_v ——volatilization rate constant;

c ——concentration of organic matter in water, $g \cdot L^{-1}$;

t ——volatilization time, s.

According to equation (4-4-5), equation (4-4-6) can be obtained.

$$\ln \frac{c_0}{c} = K_v t \quad (4\text{-}4\text{-}6)$$

Therefore, the time ($t_{1/2}$) required for the volatilization of half of the organic substances can be obtained by equation (4-4-7).

$$t_{1/2} = \frac{0.693}{K_v} \quad (4\text{-}4\text{-}7)$$

If L is the height of solution in a container with a certain section surface, the relationship between mass transfer coefficient and volatilization rate constant K_v is shown as equation (4-4-8).

$$K_v = \frac{K}{L} \quad (4\text{-}4\text{-}8)$$

Therefore, as long as the volatilization rate constant K_v of a compound is obtained, the mass transfer coefficient K can be obtained.

3 Experimental instruments and reagents

3.1 Instruments

UV spectrophotometer, electronic balance, weighing flask, beaker, volumetric flask, ruler.

3.2 Reagents

Toluene and methanol (both analytical pure).

4 Experimental procedure

(1) Preparation of reserve solution: 2.500g toluene is accurately weighed and placed in a 250mL volumetric flask. Methanol is used to dilute the solution to the scale, and the solution concentration is about 10mg/mL.

(2) Preparation of intermediate solution: Take 5mL the above reserve solution and place it in a 250mL volumetric flask, and dilute it with water to the scale. The solution concentration is 200mg/L.

(3) Drawing of standard curve: Take 0.25, 0.5, 1.0, 1.5 and 2.0mL toluene intermediate solution respectively in 10mL volumetric flasks and dilute them with water to the scale. The concentrations are 5, 10, 20, 30 and 40mg/L respectively. The standard curve of toluene can be obtained by measuring the absorbance at 205nm with UV spectrophotometer and plotting the mass concentration with the absorbance.

(4) Pour the remaining intermediate solution of toluene into two beakers, measure the solution height L, and record the time. Let it evaporate naturally, take samples with 1mL every 10min, dilute samples with water to 10mL, measure the absorbance at the wavelength of 205nm, and measure 10 points in total.

5 Data processing

5.1 Find the half-life ($t_{1/2}$) and the volatilization rate constant of toluene

The concentration of toluene in solution at different reaction times is obtained from the standard curve, and the relation curve of $\ln(c_0/c)$-t is drawn. Then $t_{1/2}$ could be obtained from its slope (K_v).

$$t_{1/2} = \frac{0.693}{K_v}$$

5.2 Find the mass transfer coefficient K

K of the compound can be calculated from equation (4-4-8) $K_v = \frac{K}{L}$.

6 Questions

What are the factors that affect the volatilization of organic pollutants in the environment?

Experiment 23 Simple identification of chromium in sediment

1 Experimental purpose

(1) Learn the extraction technology of chromium in different states of sediment.

(2) Determine the content of chromium in different states.

2 Experimental principle

Chromium pollution in sediment of water is mostly caused by sewage discharge from electroplating, tanning and dyeing. In reductive and neutral water, hexavalent chromium discharged from pollution sources can be easily reduced to trivalent chromium. Trivalent chromium in water is mainly in the form of hydroxide. It is easy to be adsorbed on suspended particles and migrate with water. In the process of migration, it deposits gradually and enters the sediment.

Chromium in sediment is mainly composed of chromium adsorbed by inorganic and organic exchangers, chromium deposites in the form of hydroxide, chromium adsorbed by humus which can be replaced by strong acid, chromium forms stable complexes with high molecular weight humus, and compound chromium formed in the most stable structure.

Among the above main states, exchangeable chromium is exchanged by one or two valence cations, such as NH_4^+, Na^+, Ba^{2+}, and some mixed electrolyte solutions. Chromium exists in the form of hydroxide and adsorbed can be dissolved in dilute acid. Chromium which forms stable complexes with high molecular weight humus, can be extracted by dilute alkali. The most stable residue is dissolved with nitric acid and perchloric acid.

In this experiment, chromium is gradually dissolved out of the sediment samples with seawater, dilute acid, dilute alkali, nitric acid perchloric acid, and then the content of each dissolved solution is measured.

3 Experimental instruments and reagents

3.1 Instruments

Model 721 (or 72) spectrophotometer, temperature-controlled electric furnace, electric centrifuge, water bath, colorimetric tube of 25mL, centrifuge tube of 50mL, conical flask of 100mL, volumetric flask of 100mL, measuring cylinders of 25mL and 50mL, pipettes of 2mL, 5mL and 10mL.

3.2 Reagents

Saturated NaOH solution, NaOH solution of 0.1mol/L, HNO_3 solution of 0.1mol/L, H_2SO_4 solution of 0.5mol/L ($V/V=1:1$), 4% $KMnO_4$ solution, 95% ethanol, natural seawater (or simulated seawater, containing 3.3% NaCl), analytical grade concentrated HNO_3, analytical grade concentrated $HClO_4$, 0.1% (mass concentration) methyl orange indicator, Cr(Ⅵ) storage solution of 160μg/mL, Cr(Ⅵ) standard solution of 4μg/mL.

Chromogenic agent: 0.2g diphenyl carbonyl dihydrazide is dissolved in 50mL acetone. Dilute it with water to 100mL and shake well. It is stored in a brown bottle and refrigerator. It cannot be used after the color darkens.

Bottom mud: After drying, pass it through 150mesh screen.

4 Experimental procedure

Weigh two about 0.5g (accurate to 0.0001g) sediment and place them in two centrifugal tubes of similar weight.

4.1 Exchangeable chromium

Add 20mL natural seawater to each tube and stir for half an hour. Centrifuge the centrifuge tubes in a symmetrical position for ten minutes, pour the supernatant into a 100mL conical flask, then add 20mL distilled water into the centrifuge tube, stir evenly and centrifuge for 10minutes, then combine the supernatant into a 100mL conical flask. Mud samples in centrifugal tubes are provided for the following experiment.

4.2 Acid soluble chromium

Add 20mL 0.1mol/L nitric acid solution into the centrifuge tubes and stir for half an hour. Centrifuge for ten minutes, pour the supernatant into a 100mL conical flask, then wash the mud sample with 20mL distilled water, centrifuge and separate the supernatant and merge it into the acid solution. Mud samples in centrifugal tubes are provided for the following experiment.

4.3 Alkali soluble chromium

Add 20mL 0.1mol/L sodium hydroxide solution into the centrifuge tubes, and put the centrifuge tubes in a water bath at 80℃ and stir for half an hour. Centrifuge for ten minutes, pour the supernatant into a 100mL conical flask, and then wash the mud sample with 20mL distilled water. Centrifuge and separate the washing solution and then add it to the alkali solution. Mud samples will be used for the following experiment.

4.4 Residual state chromium

Mud samples in the centrifuge tubes are quantitatively washed into 100mL conical flasks with 40mL distilled water.

4.5 Digestion process

Add 1mL concentrated nitric acid to each conical flask, and then put all conical flasks on the temperature-controlled electric furnace to concentrate the solution to about 10mL. Add 4mL perchloric acid into each flask, cover with the surface dish, and continue to digest until the water sample is clear, the mud sample is white and the bottle is filled with white smoke. If residue is not white, add 4mL concentrated nitric acid and continue to digest until residue becomes white. Remove the conical flasks, cool down and transfer the digestion solution into 100mL volumetric flasks. Dilute them to 100mL with distilled water. Remove 5mL solution from each bottle into a clean 100mL conical flask, add 15mL distilled water, mark each bottle, and add glass beads. Then take out a bottle of solution, add a drop of methyl orange indicator, add saturated sodium hydroxide solution drop by drop until the solution turns yellow, then slowly add 0.5mol/L sulfuric acid until it turns red. In this way, the pH values of the solution in eight flasks are adjusted one by one. Then add a drop of 1:1 sulfuric acid respectively.

4.6 Oxidation treatment

Adjust the volume of the above eight flasks of solution to about 25mL with distilled water, then heat them on the electric stove, and add two drops of 4‰ $KMnO_4$ solution after boiling. Boil for another two minutes, then add 1mL ethanol along the bottle wall while it is still hot. Boil for another two minutes. Remove the conical flasks, cool down and set the pH value of each solution to neutral (indicated by the pH test paper), then filter them into 25mL colorimetric tubes, wash the conical flasks and funnels with distilled water, and make the filtrate in the tube is exactly 25mL.

4.7 Preparation of standard solution

Take six 25mL colorimetric tubes, add 0, 0.5, 1.0, 1.5, 2.0 and 2.5mL Cr(Ⅵ) standard solution respectively, and dilute them to the scale with distilled water.

4.8 Determination

Each of the six colorimetric tubes shall be one batch. Add 1.25mL chromogenic agent respectively and shake well. Color development is for ten minutes. The absorbance of each chromogenic solution is determined with reference of standard solution No. 0 at 540nm wavelength on a spectrophotometer with a 2cm covette. After measuring, wash the glass instruments, and then soak them with nitric acid lotion.

5 Data processing

(1) Take mass concentration of Cr(Ⅵ) in the standard solution as the x-coordinate and absorbance as the y-coordinate to make the standard curve.

(2) Mass concentration of Cr(Ⅵ) is obtained according to the standard curve of absorbance of chromochromic solution in each state, and then the mass of chromium in each kilogram of mud sample is further calculated, and their average value is finally calculated. Fill data in the Table 4-4-2 and Table 4-4-3.

Table 4-4-2 Standard curve data

Mass of Cr(Ⅵ)/μg						
Absorbance						

Table 4-4-3 Status and content data

Projects	Exchange state		Acid soluble state		Alkali soluble state		Residue state	
	1	2	1	2	1	2	1	2
Absorbance								
Mass of Cr(Ⅵ)/μg								
Mass concentration of chromium/(μg/mL)								
Average mass concentration of chromium/(μg/mL)								

Notes: Digestion and oxidation must be carried out in the fume hood, and pay attention not to make strong acid and strong alkali solution splash.

6 Questions

(1) Think about what states of chromium exist in sediment and explain reasons.

(2) Indicate the main forms of chromium in sediment according to the experimental results.

Experiment 24 Study on the existence forms and migration rules of heavy metals in sediment

1 Experimental purpose

(1) Understand the existing forms and migration rules of heavy metals in sediment.

(2) Master the chemical extraction and operation technologies of heavy metals in sediment with different existing forms.

(3) Master the basic principles and correct operation skills of atomic absorption method.

2 Experimental principle

2.1 The types of existing forms

To analyze the migration and transformation rules of pollutants in water, we must first understand the forms of pollutants in water and the relationship between the existing forms, and the study of heavy metal pollutants is no exception. Tang Hongxiao put forward that "the so-called form actually includes four aspects: valence state, chemical state, combined state and structural state, which may show different biological toxicity and environmental behavior". The existing form analyzed here mainly refers to the combined state of heavy metals in water. The existing forms of heavy metals in water can be divided into dissolved state and granular state, that is, the $0.45\mu m$ filter membrane is used to filter the water sample. Metals dissolved in filtered water is in the dissolved state, and unfiltered metals in the original water sample is in the granular state (including the suspended state in the suspended substances and the deposited state in the surface sediments). By using the step-by-step chemical extraction method proposed by Tessier et al., granular heavy metals can be further divided into the following five existing forms. The first is exchangeable state, which refers to the heavy metals adsorbed on the surface of clay, minerals, organic matter or iron and manganese hydroxide in suspended sediments. The second is the carbonate combined state, which refers to the heavy metals bound on the carbonate precipitation. The third is the combined state of iron and manganese hydrated oxide, which refers to the formation and combination of heavy metals with hydrated iron oxide and manganese oxide in water. The fourth is the organic sulfide and sulfide combined state, which refers to that heavy metals in the particulate matter enter or wrap on the organic particles in different forms, and react with the organic matter or generate sulfide. The residual state refers to the part of the heavy metals present in the crystal lattice of quartz, clay, minerals and other crystalline minerals.

2.2 Transfer property

The migration properties of different forms of heavy metals in water are different. Dissolved heavy metals have the most direct impact on human beings and aquatic ecosystems, and are one of the common bases for judging the degree of heavy metal pollution in water. The composition of granular heavy metals is complex and their morphology and properties are different. Exchangeable heavy metals are the most unstable, and as long as environmental conditions change, they are easily dissolved in water or exchanged by other ions with strong polarity, which is an important part of affecting water quality. Carbonate bound heavy metals are most easily re-released into water when the environment changes, especially when pH value changes. Iron and manganese hydrated oxide binding heavy metals are also partially released when the environment changes. Organic sulfides and sulfide binding heavy metals are not easily absorbed by organisms and are stable. Residual heavy metals are the most stable and will not be released into the water for a long time.

2.3 Research methods of migration rules

Different forms of heavy metals have different chemical conditions and difficulties in separating from the combined carrier, that is, there are differences in stability, so the pollution degree to the water is also different. Different heavy metal pollutants have different distribution patterns of different forms in water. The migration and transformation process of heavy metals can be studied by studying their distribution differences and mutual transformation process, which can be used as the basis for judging their harm to water. When analyzing heavy metal pollution in sediments, it is not enough to only recognize the total amount of heavy metals, but also to analyze the content and distribution law of each component. Then discuss the pollution nature, transformation mechanism and potential pollution to water of heavy metal pollutants in sediments. In the study of heavy metal pollution in water, this method is often used to analyze the main forms of mutual migration of heavy metals in the water phase and the solid phase, so as to obtain the dynamic transformation rules and the final destination of different forms of heavy metals in water.

3 Experimental instruments and reagents

3.1 Instruments

3200 atomic absorption spectrophotometer with hollow cathode lamp (copper, lead, cadmium and zinc), electric centrifuge and plastic centrifuge tube, temperature control magnetic stirrer and magnetic stirring bar, quartz beaker (50mL), volumetric flask (50mL, 25mL), pipette (25mL, 10mL, 5mL, and 2mL), PTFE beaker (50mL), plastic measuring cylinder (25mL, 10mL), measuring cylinder (50mL, 10mL), electric heating plate, mussel type mud sampler, magnetic mortar, agate mortar, nylon screen (100mesh), and redistilled water.

3.2 Reagents

1mol/L $MgCl_2$ (adjust pH to 7 with HCl), 1mol/L NaAc (adjust pH to 5 with

HAc), 0.02mol/L HNO_3, concentrated HF (GR), concentrated $HClO_4$ (GR), 0.04mol/L $NH_2OH \cdot HCl$ (dissolved in 25% HAc), 30% H_2O_2 (adjust pH to 2 with HNO_3), 3.2mol/L NH_4Ac (dissolved in 20% HNO_3).

Standard reserve solution of Cu, Pb, Cd, Zn (1000mg/L).

Mixed standard solution: Metal standard reserve solution is diluted with 0.2% nitric acid to make the mixed standard solution containing 50.0, 100.0, 10.0 and 10.0μg of Cu, Pb, Cd and Zn, respectively.

4 Experimental procedure

4.1 Sampling and sample pretreatment

Grab the surface sediment with a mussel type mud sampler (or directly with other containers), immediately remove the surface layer with a stainless steel knife, take hundreds of grams of samples from the center, seal them in a plastic bag, and store them in a refrigerator.

Before use, take out the mud sample from the plastic bag, put it in the glass surface dish or petri dish, and dry it in the non-polluting air circulation. The dry sample is firstly porphyrized in the magnetic mortar, and then ground to 80-100mesh with the agate mortar, and stored in the dryer for later use.

4.2 Chemical continuous extraction method for the heavy metals in different states in sediment

4.2.1 Exchangeable state (adsorbed state)

Accurately weigh 1.000g sample in a quartz beaker, add 25mL 1mol/L $MgCl_2$ and stir for one hour, then centrifuge and settle. The centrifugate is transferred into a 50mL volumetric flask, and the residue is washed with 10mL distilled water. The centrifugate is incorporated into a 50mL volumetric flask and diluted to scale for the determination of exchangeable heavy metal content. Reserve residue 1 is retained for extraction of carbonate combined heavy metals.

4.2.2 Carbonate combined state

The residue 1 is transferred into a quartz beaker with 20mL 1mol/L NaAc (adjust pH = 5.00 with HAc in advance) for several times, stirred and extracted for five hours, and centrifuged. The centrifugate is transferred into a 50mL volumetric flask, the residue is washed with 10mL redistilled water, and the centrifugate is centrifuged and precipitated. The centrifugate is incorporated into a 50mL volumetric flask and diluted to a scale with redistilled water for the determination of the content of the carbonate combined heavy metals. The residue 2 is reserved for extracting the iron and manganese hydrated oxide combined heavy metals.

4.2.3 Iron and manganese hydrated oxide combined state

The residue 2 is transferred into the quartz beaker in stages with 40mL 0.04mol/L $NH_2OH \cdot HCl$ (dissolved in HAc with 25% volume fraction), stirred and extracted in water bath at (96±3)℃ for five hours, and centrifuged. Centrifugate is transferred into a

50mL volumetric flask, and the residue is washed with 10mL redistilled water and centrifuged for settlement. The centrifugate is diluted to scale with water in a 50mL volumetric flask, and the solution is used for the determination of the content of iron and manganese hydrated oxide combined heavy metals. The residue 3 is retained for the extraction of organic sulfide and sulfide combined heavy metals.

4.2.4 Organic sulfide and sulfide combined state

The residue 3 is transferred into a quartz beaker with 6mL 0.02mol/L HNO_3 and added 10mL 30% H_2O_2 (adjust pH=2 in advance with HNO_3), stirred and extracted in (85±2)℃ water bath for two hours, continue to be stirred for three hours, took out and cooled, then added 10mL 3.2mol/L NH_4AC (dissolved in HNO_3 with 20% volume fraction) and 15mL redistilled water, stirred for one hour, and centrifuged for sedimentation. The centrifugate is transferred to a 50mL volumetric flask, and the residue is washed with 10mL redistilled water. The centrifugate is incorporated into a 50mL volumetric flask, and diluted to scale with redistilled water. The solution is used for the determination of the content of organic sulfide and sulfide combined heavy metals. Reserve residue 4 for digestion.

4.2.5 Residual state

The residue 4 is transferred into a PTFE (50mL) beaker in stages with 15mL $HClO_4$. Add 5mL HF to the solution (measured with plastic measuring cylinder), and heat and digest it (pay attention not to evaporate completely) with electric heating plate (or electric stove with asbestos net) in the fume cupboard. The above proportions of $HClO_4$ and HF are repeated until the sediment is completely digested and evaporated. Add 1mL $HClO_4$ and a small amount of redistilled water (10-20mL) and heat slightly. The solution should be clear. The liquid is transferred into a 50mL volumetric flask. Wash the PTFE beaker with a little slightly acid ($HClO_4$ acidification) distilled water for several times, and transfer it to the volumetric flask, where it is diluted to the scale. The obtained solution can be used for the determination of the content of residual heavy metals.

4.3 Determination

4.3.1 Preparation of standard solution

The mixed standard solution of 0, 0.25, 0.50, 1.50, 2.50 and 5.00mL are absorbed respectively into six 50mL volumetric flasks, and diluted with 0.2% HNO_3 to constant volume. The mass concentration of each metal in this mixed standard series solution is shown in Table 4-4-4.

Table 4-4-4 Preparation and concentration of mixed standard series solution

Mixed standard solution volume/mL		0	0.25	0.50	1.50	2.50	5.00
Metal mass concentration in mixed standard series solution/(mg/L)	Cu	0	0.25	0.50	1.50	2.50	5.00
	Pb	0	0.50	1.00	3.00	5.00	10.0
	Cd	0	0.05	0.10	0.30	0.50	1.00
	Zn	0	0.05	0.10	0.30	0.50	1.00

Notes: Constant volume is 50mL.

4.3.2 Determination of standard sample and sample solution

The standard series solution and the sample solution to be tested are respectively determined by atomic absorption spectrometry under the same conditions. The determination conditions of atomic absorption are shown in Table 4-4-5. The obtained absorbance value is plotted against the concentration and the working curve is made. The mass concentration of the element to be measured is obtained from the curve.

Table 4-4-5　Atomic absorption determination conditions

Element	Cu	Pb	Cd	Zn
Wavelength/nm	324.7	283.3	228.8	213.8
Lamp current/mA	10	10	8	10
Slit location	2	2	2	2
Damping position	1	1	1	1
Gain position	2.9	4.0	4.5	4.5
Fuel gas	Acetylene	Acetylene	Acetylene	Acetylene
Oxidant gas	Air	Air	Air	Air
Flame type	Oxidation type	Oxidation type	Oxidation type	Oxidation type

5　Data processing

mass fraction (mg/kg) of copper, lead, cadmium and zinc in sediment

$$= \frac{MV}{W} \tag{4-4-9}$$

Where, M——mass concentration of the sample detected on the standard curve, mg/L;

V——final constant volume of copper, lead, cadmium and zinc, mL;

W——mass of the sediment, g.

6　Questions

(1) Why is it necessary to conduct morphological analysis of sediment?

(2) Why pay attention to the pH of the extractant?

(3) What should be paid special attention to in this experiment?

(4) According to the experimental results, the pollution status and behavior of Cu, Pb, Cd and Zn in sediment are evaluated.

(5) What conditions should be tested for atomic absorption spectrometry?

7　Notes

(1) Because the sources of sediment are different, the content of copper, lead, cadmium and zinc is different, so the standard curve can be adjusted appropriately according to the specific situation.

(2) When HF is used to decompose residue, heating should not exceed 250℃, so as to avoid the decomposition of PTFE and produce toxic gas containing fluoroisobuty-

lene. Samples are best decomposed in platinum vessels. In addition, HF is toxic and corrosive to human body. When using it, please pay attention not to inhale HF steam or touch HF. HF may cause burns and ulceration after contacted with skin, and it is not easy to heal.

(3) When measuring samples, different blank control tests should be made for different forms.

Experiment 25 Determination of total arsenic content in food by hydride atomic fluorescence spectrometry

Arsenic is a natural composition in the earth's crust, which is widely distributed in the atmosphere, water and land environment. Arsenic steam has a smelly garlic smell. Arsenic can easily react with fluorine and oxygen, and will react with most of the metals and nonmetals under the condition of heating. Arsenic is insoluble in water, but soluble in nitric acid, aqua regia, and strong alkali solution.

Arsenic trioxide is highly toxic. After entering the human body, it will destroy certain cellular respiratory enzymes, making tissue cells unable to obtain oxygen and die. In addition, it can strongly damage the gastrointestinal mucosa, causing mucosal ulceration and bleeding. And it can damage blood vessels and liver, cause bleeding, and even cause respiratory and circulatory failure and death.

The biggest arsenic threat to public health comes from contaminated groundwater. In industry, arsenic is used as an alloy additive, as well as in the treatment of glass, coatings, textiles, paper, metal adhesives, wood preservatives, and ammunition. Arsenic is also used in tanning processes, insecticides, feed additives, and pharmaceuticals to a limited extent. Smoking can expose the human body to the natural inorganic arsenic contained in tobacco, which is mainly the naturally occurring arsenic ingested by tobacco plants from the soil. The early symptoms of acute arsenic poisoning include vomiting, abdominal pain and diarrhea, followed by paralysis and tingling of limbs, muscle spasm, and death in extreme cases.

Long term exposure to arsenic in drinking water and food can lead to cancer and skin lesions. Arsenic is one of the 10 chemicals listed by the World Health Organization as causing significant public health concern. The current recommended limit for arsenic content in drinking water is $10\mu g/L$. Therefore, mastering the determination of total arsenic content in food by hydride atomic fluorescence spectrometry is of great significance.

1 Experimental purpose

(1) Master the basic principle and operational steps of determining total arsenic content in food using hydride atomic fluorescence spectrophotometry.

(2) Understand the impact of arsenic on human health.

2 Experimental principle

Thiourea is added to the food sample after wet digestion or dry ashing to reduce pentavalent arsenic to trivalent arsenic, and then sodium borohydride is added to reduce it to arsine. Arsine loaded by argon decomposes into atomic arsenic into the quartz atomizer, and generates atomic fluorescence under the excitation of the emission light of the specially made arsenic hollow cathode lamp. Its fluorescence intensity is proportional to the arsenic concentration in the measured solution under fixed conditions, and is quantitative compared with the standard series.

3 Experimental instruments and reagents

3.1 Instruments

Atomic fluorescence spectrophotometer.

3.2 Reagents

(1) Sodium hydroxide solution (2g/L, 100g/L), thiourea solution (50g/L), sulfuric acid solution (1+9).

(2) Sodium borohydride solution (10g/L): Weigh 10.0g sodium borohydride, dissolve it in 1000mL 2g/L sodium hydroxide solution, and mix well.

(3) Arsenic standard stock solution (containing arsenic 0.1g/L): Accurately weigh 0.1320g arsenic trioxide dried at 100℃ for more than 2h, dissolve it with 100g/L sodium hydroxide solution, transfer it into a 1000mL volumetric flask, add 25mL sulfuric acid (1+9), and fix the volume.

(4) Arsenic standard solution (containing arsenic 1μg/mL): Suck 1.00mL arsenic standard stock solution into a 100mL volumetric flask and dilute it with water to the scale. This liquid should be prepared and used on the same day.

(5) Wet digestion reagents: Nitric acid, sulfuric acid, perchloric acid.

(6) Dry ashing reagents: Magnesium nitrate hexahydrate (150g/L), magnesium chloride, hydrochloric acid (1+1).

(7) Arsenic standard series solution: Take six 25mL volumetric flasks or colorimetric tubes, accurately add 0, 0.05, 0.2, 0.5, 2.0, and 5.0mL 1μg/mL arsenic standard solution, add 12.5mL sulfuric acid (1+9) and 2.5mL 50g/L thiourea respectively, add water to the scale, and mix well for testing. Each equivalent to arsenic concentration of 0, 2.0, 8.0, 20.0, 80.0, and 200.0ng/mL.

4 Experimental procedure

4.1 Sample digestion

(1) Wet digestion: Weigh 1-2.5g solid sample and 5-10g (or mL) liquid sample (accurate to the second place after the decimal separator), put them into a 50-100mL conical flask, and make two reagent blanks at the same time. Add 20-40mL nitric acid and 1.25mL sulfuric acid, shake well, and place them overnight. Heat and digest samples on an electric

heating plate. If there are still undecomposed substances or darkened colors when the digestion solution is treated to about 10mL, cool it down, add 5-10mL nitric acid, and then digest it to about 10mL for observation. Repeat this process two or three times, and pay attention to avoiding carbonization. If the digestion is still incomplete, add 1-2mL perchloric acid, continue to heat until the digestion is complete, then continue to evaporate until the white smoke of perchloric acid is exhausted, and the white smoke of sulfuric acid begins to emit. Cool and add 25mL water, then evaporate until sulfuric acid white smoke appears. Cool it down, transfer the contents into a 25mL volumetric flask or colorimetric tube with water, add 2.5mL 50g/L thiourea solution, add water to the scale, mix well, and prepare for testing.

(2) Dry ashing: Weigh 1-2.5g (accurate to the second place after the decimal separator) solid sample into a 50-100mL crucible, and make two reagent blanks at the same time. Add 10mL 150g/L magnesium nitrate solution, mix well, and evaporate to dryness at low heat. Carefully cover 1g magnesium oxide on the slag and carbonize it on the electric furnace until there is no black smoke. Then transfer it to a high-temperature ashing furnace at 550℃ for 4hours. Take it out for cooling, and carefully add 10mL hydrochloric acid (1+1) to neutralize magnesium oxide and dissolve the ash. Transfer it into a 25mL volumetric flask or colorimetric tube, add 2.5mL 50g/L thiourea solution, brush the crucible with sulfuric acid (1+9) for several times, then transfer it into a volumetric flask or colorimetric tube until it reaches the 25mL scale, and mix well for testing.

4.2 Determination

(1) Instrument reference conditions: Photomultiplier tube voltage: 400 V; Arsenic hollow cathode lamp current: 35mA; Atomizer: temperature is 820-850℃, height is 7mm; Argon flow rate: 600mL/min; Measurement method: direct reading of fluorescence intensity or concentration; Reading method: peak area; Reading delay time: 1 s; Reading time: 15 s; Adding time of sodium borohydride solution: 5 s; Volume of standard solution or sample solution added: 2mL.

(2) Concentration measurement method: direct measurement of fluorescence intensity. After turning on and setting the instrument conditions, preheat for about 20minutes, press the "B" button to enter the blank value measurement state, continuously inject samples with the standard series "0" tube, and wait for the reading to stabilize. Press the neutral button to record the blank value (that is, let the instrument automatically deduct background influence) to start the measurement. Measure the standard series in sequence (at this point, the "0" tube can not be measured). After the standard series is tested, the injector should be carefully cleaned and the measurement data should be recorded (or printed).

(3) Instrument automatic mode: Use the software function provided by the instrument for concentration direct reading measurement. After starting up, setting conditions, and preheating, necessary parameters need to be input, that is, sample amount (g or mL), dilution volume (mL), injection volume (mL), the concentration unit of the result, the

number of repeated measurements at each point of the standard series, the number of points of the standard series (excluding zero points) and the concentration values of each point. Firstly, enter the blank value measurement state, continuously inject samples with the "0" tube of the standard series to obtain a stable blank value, and then automatically deduct background. Then, sequentially measure the standard series (at this time, the "0" tube needs to be measured again). Before measuring the sample solution, enter the blank value measurement state again. After testing with the standard series "0" tube to restore and stabilize the reading, two reagent blanks are injected separately, and the instrument takes their average value as the blank value for deduction. Subsequently, the samples can be tested sequentially. After the measurement is completed, return to the main menu and select "Print Report" to print the measurement results.

5 Data processing

If fluorescence intensity measurement is used, the results of the standard series need to be regressed first (as the "0" tube is forced to be 0 during measurement, the zero point value should be input to occupy a point position). Then, the arsenic mass concentration of the reagent blank solution and the tested solution of the sample should be calculated based on the regression equation, and the arsenic content of the sample should be calculated according to equation (4-4-10):

$$x = \frac{c_1 - c_0}{m} \times \frac{25\,\text{mL}}{1000} \qquad (4\text{-}4\text{-}10)$$

Where, x——arsenic content of the sample, mg/kg or mg/L;

c_1——mass concentration of the tested sample solution, ng/mL;

c_0——mass concentration of reagent blank solution, ng/mL;

m——mass or volume of the sample, g or mL.

Two significant digits are reserved for the calculation results.

6 Questions

(1) Analyze and discuss the main sources of arsenic in food.

(2) What are the effects of arsenic on human health?

References

[1] GONG A J, LIU J M, WANG H O. Chemistry of the environment (bilingual)[M]. Beijing: Chemical Industry Press, 2015.

[2] DONG D M, HUA X Y, KANG C L. Environmental chemistry experiment[M]. Beijing: Peking University Press, 2010.

[3] GU X Y, MAO L. Environmental chemistry experiment[M]. Nanjing: Nanjing University Press, 2012.

[4] WU C Q, SUN H, DENG H M. Environmental integrated chemistry experiment course[M]. Beijing: Beijing Institute of Technology Press, 2019.

[5] HJ 479—2009 Ambient air—Determination of nitrogen oxides—N-(1-Naphthyl)ethylene diamine dihydrochloride spectrophotometric method.

[6] HJ 504—2009 Ambient air—Determination of ozone—Indigo disulphonate spectrophotometry.

[7] GB/T 7489—1987 Water quality—Determination of dissolved oxygen—Iodometric method.

[8] GB/T 13193—1991 Water quality—Determination of TOC by nondispersive infrared absorption method.